M & E HANDBOOKS

M & E Handbooks are recommended reading for examination syllabuses all over the world. Because each Handbook covers its subject clearly and concisely books in the series form a vital part of many college, university, school and home study courses.

Handbooks contain detailed information stripped of unnecessary padding, making each title a comprehensive self-tuition course. These are amplified with numerous self-testing questions in the form of Progress Tests at the end of each chapter, each text-referenced for easy checking. Every Handbook closes with an appendix which advises on examination technique. For all these reasons, Handbooks are ideal for pre-examination revision.

The handy pocket-book size and competitive price make Handbooks the perfect choice for anyone who wants to grasp the essentials of a subject quickly and easily.

THE M. & E. HANDBOOK SERIES

BASIC BIOLOGY

P. T. MARSHALL, M.A.

Senior Biology Master,
The Leys School, Cambridge

MACDONALD AND EVANS

MACDONALD AND EVANS LTD.
Estover, Plymouth PL6 7PZ

First published 1974
Reprinted 1976
Reprinted in this format 1978

©

MACDONALD AND EVANS LIMITED
1974

ISBN: 0 7121 0232 9

*Printed in Great Britain by
Richard Clay (The Chaucer Press), Ltd.,
Bungay, Suffolk*

PREFACE

THIS HANDBOOK has been written to act as an elementary but companion volume to the author's *Biology, Advanced Level*, published in this series three years ago. The method used to present information is standard to the HANDBOOK series and examination questions from recent "O" Level biology papers set by a number of boards have been included at the end of each chapter.

The main theme of the HANDBOOK is the elementary physiology of the green plant and the mammal and sufficient concise information on these topics is included to cover essential common ground to all G.C.E. syllabuses at "O" Level as well as much that is required by the C.S.E. syllabuses. For the latter reason the HANDBOOK has been called *Basic Biology*.

The author has been particularly pleased to include material in this book which it was not possible to fit into the more advanced text, in particular details of elementary plant anatomy. It has been the intention of the author and publisher that the two HANDBOOKS used in conjunction would be most useful to those students who take up biology late in their academic careers and require a "crash course" from no previous knowledge to advanced level grade.

While this HANDBOOK is essentially about green plants and mammals, a synopsis of the classification of other lower organisms is included in the last chapter so that students can fit into place various types whose lives interact most closely with higher forms.

Acknowledgment to the Associated Examination Board, the Oxford and Cambridge Examination Board and the University of Cambridge Local Examination Board is made for permission to use question material from past papers. The source of a question is indicated by the capital letter of the appropriate board and descriptive as well as multiple choice questions are included. Acknowledgment is made to the Cambridge University Press for permission to use material from *Tropical Health Science*, by D. T. D. Hughes and P. T. Marshall, 1967. Acknowledgment is also made to Peter Redmond, the general editor of this series, who suggested the writing of these two HANDBOOKS, and to the publishers.

Finally it should be noted that the HANDBOOK series is aimed at provision of concise and accurate factual information to assist examination candidates in learning and revising material. It is out of place to include practical exercises in such a book but it should be quite clear to students that modern biology is an experimental rather than a descriptive science.

November, 1973   P. T. M.

CONTENTS

LIST OF ILLUSTRATIONS

CHAPTER I

THE LIVING ORGANISM AND THE CELL

CHARACTERS OF LIVING ORGANISMS

1. The characters of living organisms. Biology is the study of living organisms and these all have certain characters, some of which are unique and some which they share with non-living entities such as machines. The characters of living organisms are set out in 2–8 below.

2. Nutrition. This is the process by which food is taken into the organism or, as with green plants, manufactured within its tissues. The food naturally consists of the same type of chemical substances out of which all living things are made, that is, carbohydrates, fats, proteins, vitamins, mineral salts and water.

(a) *Animals* take complex food molecules into their bodies and break them down within the alimentary canal or gut by a process of *digestion* (*see* Chapter V). The products of digestion are simply small molecules which are able to pass across the gut wall and thus become incorporated into the animals' own tissues.

(b) *Plants* differ from animals in their ability to manufacture food substances from carbon dioxide and water together with simple salts. This is done using light energy and involves the green pigment chlorophyll, the process being termed *photosynthesis* (*see* Chapter IV).

(c) *Saprophytes* are distinct from other organisms in their mode of nutrition in that they take in soluble organic matter directly through the body surface. The source of this matter is dead plant and animal matter broken down by the saprophytes outside their own bodies. The majority of fungi and bacteria are saprophytes (*see* Chapter III).

3. Respiration. In everyday (*i.e.* non-scientific) usage the word *respiration* is often understood to be the same as breathing but in its biological sense respiration means energy release within the tissues of an organism. The release of energy is due to a series of chemical reactions which reduce the size of the particular "fuel" molecules which are being utilised. While these energy-providing substances are being down-graded, their stores of energy are transferred to special energy-store molecules within the organism. These are then available for use by the organism as required.

The most common supplier of energy to living systems is glucose and the normal energy-store substance is adenosine tri-phosphate or ATP.

(*a*) *Aerobic respiration* involves the use of oxygen and the end-products of the reaction are carbon dioxide and water. In a simple form it may be represented thus:

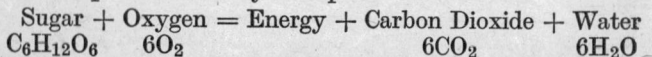

$$\text{Sugar} + \text{Oxygen} = \text{Energy} + \text{Carbon Dioxide} + \text{Water}$$
$$C_6H_{12}O_6 \quad 6O_2 \qquad\qquad 6CO_2 \qquad\quad 6H_2O$$

The process is discussed in further detail in Chapter IV but it should be realised that the equation given represents only the initial (in this case glucose) and final products of aerobic respiration.

Aerobic respiration is approximately 54 per cent efficient as compared with 4 per cent efficiency of anaerobic respiration (*see* below). It is the normal form of respiration for most living organisms although some, such as green plants, can withstand temporary anaerobic spells. This is dealt with in the first part of Chapter VI.

(*b*) *Anaerobic respiration* is the chemical breakdown of a respiratory substance without the involvement of oxygen. Again using glucose sugar as the usual substance involved, the process may be represented as follows:

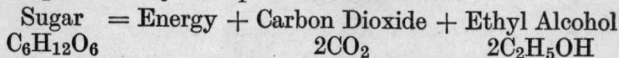

$$\text{Sugar} = \text{Energy} + \text{Carbon Dioxide} + \text{Ethyl Alcohol}$$
$$C_6H_{12}O_6 \qquad\qquad 2CO_2 \qquad\quad 2C_2H_5OH$$

It may be appreciated that alcohol is itself an inflammable liquid above certain concentrations and this clearly shows that anaerobic respiration is very inefficient and that a lot of the original energy of the sugar is being wasted.

The process is however extremely important to the survival of those organisms such as saprophytes (*i.e.* most bacteria and fungi) which live in places deficient in oxygen, usually amongst decaying matter.

Yeast is such an organism, and its anaerobic respiration is used commercially in brewing and baking.

4. Excretion. During the various chemical reactions that are involved in being alive, which run into hundreds of thousands for any living organism, waste products are produced. The formation and elimination of these is called *excretion* and this is a third activity of the living organism. The topic is covered fully in Chapter VII.

As seen above in 3, the process of respiration with final products of water, carbon dioxide and heat (or carbon dioxide, alcohol and heat in anaerobic reactions) is a major source of excretory products. There are certain differences between the excretory processes of animals and plants:

(a) *Animals* tend to take in an excess of nitrogen containing protein in their food and from this and from the breakdown of their own tissues come the nitrogen-containing excretory products. These tend to be poisonous, or toxic, to the organism and include such chemicals as urea, ammonia and uric acid.

Any excess mineral salts taken in with the food are also excreted as the bodies of animals have fixed internal concentrations of salts and upsetting the balance of these would lead to death.

At first, it may appear that animal faeces is a good example of an excretory product but in fact the bulk of waste from the gut consists of the dead bodies of bacteria as well as indigestible matter (*e.g.* cellulose from humans). The bile salts that colour faeces may be looked on as partly excretory, as may the remains of gut lining cells, but in physiology true excretory products must have come from reactions within the tissues of the organism itself.

Water may be an excretory product and is normally the medium in which other wastes are dissolved. Water, on the other hand, is also an essential dietary requirement, not least because of the role it plays in excretion.

(b) *Plants* excrete oxygen as a waste from photosynthesis. They do not have the same type of nitrogen products as animals as they manufacture their own amino acids and proteins from simple salts and other molecules. Thus plants have nothing equivalent to animal urine.

Excess of salts may be taken in by some plants under certain conditions and become stored in crystalline form within their tissues. Crystals of calcium oxalate are fairly common as excretory storage products in leaves.

Water and carbon dioxide from respiration are true plant excretory products but both these substances are utilised in photosynthesis. Since in daylight this latter process usually greatly exceeds respiration the plant will actually be taking in carbon dioxide and water.

Plants do evaporate water from their leaves in a process called *transpiration* but as only a minute fraction of such water has come from chemical reactions within their tissues it cannot be regarded as excretory.

NOTE: Excretory products tend to be small molecules which contain little energy and many organic excretory substances are toxic. In the case of some parasitic bacteria such as botulinus the excretory wastes are very poisonous indeed, a few thousandths of a gramme being sufficient to kill a man.

5. Response. Animals and plants respond to changes in their surroundings in a variety of ways which may be thought of as tending to increase their chances of survival. While the rapid nature of co-ordination and response is clear enough in the case of animals it is much less evident in plants. All living things do, however, respond to stimuli and this is a fourth characteristic feature (*see* Chapter IX).

(*a*) *Animals* tend to respond to changes in their environment by some sort of movement either of a part or of the whole of their bodies. Slow responses to general environmental changes, such as decrease in temperature or hours of daylight, are co-ordinated chemically by hormones. Faster responses due to immediate changes detected by the animal's sense organs are brought about by nerve impulses. These may travel at over 100 metres per second in a mammal and the resulting movements be extremely fast. Co-ordination and response are seen at their most highly developed in the warm-blooded birds and mammals.

(*b*) *Plants* are mostly fixed, and in the case of trees a large proportion of their bulk is quite dead and inert. Growing regions may be sensitive to directional changes in forces such as light and gravity. A few plants, such as the Venus fly trap, do show very rapid responses to environmental stimuli but on the whole plant responses take place by growth processes and thus, compared with those of animals, they are exceedingly slow.

6. Movement. This activity is related to the previous one. While all living organisms show the characteristic of movement the majority of the higher plants are fixed and only move the growing parts by slow processes. This is because the food of plants comes from the soil in the form of water and mineral salts and from the air in the form of carbon dioxide. It is clearly more advantageous to a plant to be anchored securely in the earth and spread out its leaves to the sun than to move about looking for light and food! (*see* Chapter IX).

Animals on the other hand feed primarily on plants or else upon each other. They need the capacity to move the whole of their bodies to obtain constant supplies of food and to respond in other ways. Animals have systems of contracting cells called *muscles* by whose means rapid movements can be achieved.

It should be noted that plants are very divergent in size and form, which are related to their particular surroundings. Animals on the other hand have to have a much more definite body shape if they are to move efficiently. Another related point is that many plants are radially symmetrical (*i.e.* can be bisected down any longitudinal axis) while animals, such as ourselves, are bilaterally symmetrical with a single axis of symmetry and definite anterior and posterior ends.

7. Growth. Living organisms are produced by mature organisms of the same species that are much larger than themselves. This applies whether the new creature is produced by sexual or asexual reproduction although in the former case the new individuals (produced by fusion of egg and sperm) may be very tiny indeed.

A characteristic of living things is their ability to grow by the incorporation of substances from their surroundings, hence their food is not simply analogous to the petrol supply of an engine.

Most plants and animals consist of very many cells and also many different types of cells so that growth of the whole organism normally involves the development of complex tissues and organs from unspecialised cells.

The growth of all organisms has quite a lot in common. Thus
it tends, after a slow start, to take place rapidly during youth,
and level off later in life generally producing an S-shaped
curve when weight is plotted against time as shown in the
diagram (*see* below). In both plants and animals growth and
form are chemically controlled from within but also subject
in varying degrees to external influences. As stated above, this
latter is particularly true of plants.

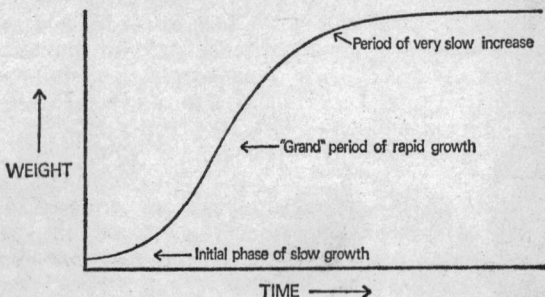

The growth curve showing: Period of very slow increase; "Grand" period of rapid growth; Initial phase of slow growth. WEIGHT plotted against TIME.

8. Reproduction. This is the last of the seven characteristics
of living organisms and perhaps the feature which distinguishes
them most clearly from all other forms of organisation such as
machines. Reproduction is an end to which all other activities
are subordinate as successful multiplication ensures survival of
a particular species while the converse leads to eventual
extinction (*see* Chapter X).

(*a*) *Sexual reproduction* is of great importance in biology
because it involves genetic mixing and interchange between
individuals which in turn leads to variation in the offspring.
On this variation the forces that bring about evolution can
act.

By definition sexual reproduction entails the fusion of a
male sex cell (the sperm) with a female sex cell (the egg). As
we shall see in Chapter X these sex cells have only a single
set of characters or genes whereas all the other cells of the
organism have two sets. Thus when the sex cells fuse to produce the fertilised egg, or zygote, this will have the normal
two sets of the adult.

Because humans are clearly divided into male and female
sexes it is common to suppose this is a general rule. Cer-

tainly it is so for higher animals. The lower animals, for example the worms, and the majority of both higher and lower plants, carry both sexes in the adult and are called *hermaphrodite*. Despite this, the important aspect of sexual reproduction is out-breeding, that is fertilisation occurring between separate individuals of a species. We shall see that many hermaphrodite organisms have means of preventing or reducing self-fertilisation.

(*b*) *Asexual reproduction* is much more common in plants than in animals although it is found in many lower animals. In the form of spore production it occurs generally throughout the plant kingdom while in the form of vegetative reproduction (where a part of the parent body breaks away to form an independent organism) it is common in higher plants, *e.g.* potato.

Asexual reproduction does not produce variation in the offspring who will grow to resemble the parent in all respects. Despite the long-term disadvantage of this the process has many immediate benefits and is much less hazardous than

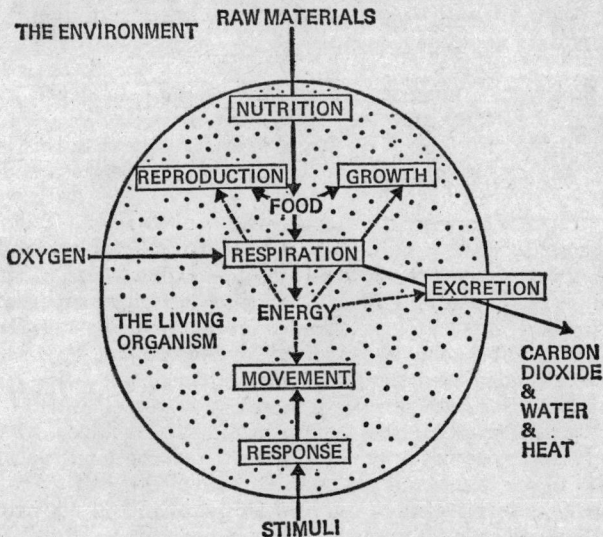

FIG. 1. Characteristic activities of the living organism.

sexual processes. It is common for plants to employ both methods of reproducing.

(c) *Nucleic acids and reproduction.* In the last two decades it has become customary to include the possession of nucleic acid as a further characteristic of all living organisms.

Nucleic acids (*see* Chapter II) are the substance of the nucleus of the cells which carry heredity information and also exist in the cytoplasm of the cell where they are instrumental in translating this information into protein manufacture.

Among their other properties the nucleic acids of the nucleus are able to replicate themselves exactly and this is an essential to the cell division involved in growth and all types of reproduction.

SOME PROBLEMS OF DEFINITION

A scheme showing the relationship between the various physiological activities of organisms is shown in Fig. 1. While the seven characters given in 2–8 above are those associated with living organisms there are certain points that should be noted at this stage:

9. Death. While it is tempting to think of death as the opposite of life and therefore as a necessary character of organisms, we find as we learn more biology that this is not true. Very many lower organisms do not seem to have the same type of life span and ageing process with which we are familiar for ourselves. After a certain time depending on such factors as food availability, temperature etc., these lower organisms (*e.g.* bacteria, many protozoa, and some algae and fungi) split into new individuals and these in turn feed and grow and multiply. Death may indeed occur by accident or starvation or being eaten or by drying up but it is not an inherent characteristic of all living organisms.

10. Living organisms and machines. Machines such as motor cars or computers may have many common characters with living things and yet no-one would claim the former to be alive. Motor cars move and feed on petrol and oil, they respire using oxygen and produce waste gases, carbon dioxide and water, while computers are built to respond to information and provide answers to problems in a way resembling that of the human mind.

Machines are however not capable of either growth or reproduction nor do they contain the essential life substance nucleic acid. There are other differences as well. The living organism depends on its chemical reactions which are catalysed by special proteins called *enzymes* (*see* Chapter II). These allow reactions to occur at low energy levels quite impossible to internal combustion engines. True enzymes will function outside the body to bring about reactions and in the last two years it has even been possible for chemists to manufacture certain enzymes.

Despite certain superficial resemblances a review of the sum total of its features usually presents little difficulty in classifying an object as living or non-living.

11. Border line organisms. There are certain entities called *viruses* that are very difficult to classify. They consist largely of nucleic acid and are able to reproduce but they do not respire themselves and also they may be crystallised and stored for many years, still retaining their reproductive properties.

It is perhaps best to leave the viruses in a special category of their own, but in so far as they have almost certainly evolved by degeneration of more complex living things they are nearer to the living than the non-living world.

In section 18 of this chapter we shall see some of the differences between plants and animals at the cell level and how these two categories may be distinguished. Once again there are living organisms that share both plant and animal features and which it is not possible to classify.

CELLS

12. Plant and animal cells. All living things consist of cells, or in some primitive organisms, of a single cell only. The cell may be regarded as the unit of construction of the organism and all cells have a number of common features (*see* Fig. 2).

13. The generalised cell. A typical cell is some ten microns in diameter ($1\mu = \frac{1}{1000}$ mm) and more or less spherical. It consists of a nucleus, and surrounding cytoplasm which is controlled by the nucleus. The cytoplasm is bounded by a membrane which regulates the passage of substances in and out of the cell. Within the cytoplasm are certain tiny systems called *organelles*. The most common of these are the mitochondria and endoplasmic reticulum.

Some of these features can be made out with simple microscopes but elaborate details of cell ultra-structure have come from work with the electron microscope which is capable of enormous magnifications.

(a) *The nucleus* is usually spherical and is bounded by a membrane (the nuclear membrane) of its own. It contains

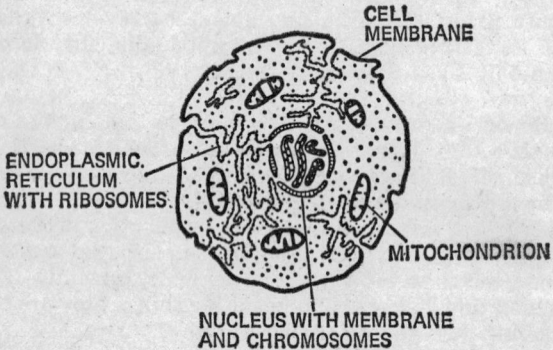

FIG. 2. Idealised structures within the cells as revealed by the electron microscope.

the nucleic acid DNA (deoxyribose nucleic acid) which is organised into the genes which determine the characters of the cell and the whole organism. The genes are themselves regions of long strands of nuclear material called *chromosomes* and these may be readily stained and observed with the light microscope. Further details of the nucleus are found in Chapter X.

(b) *The cytoplasm* was thought to be a grey amorphous jelly and indeed it still looks like this under low-power magnification. With more sophisticated instruments cytoplasm is seen to consist of a number of very highly organised structures so that the term should now be used as that part of the cell which excludes the nucleus. The word *protoplasm* which used to be given to describe all the cell contents is best discarded.

(c) *The endoplasmic reticulum* is a complex network of canals that ramify throughout the whole cytoplasm. Along the sides of the canals lie many thousands of minute bodies called *ribosomes*. These are also formed from nucleic acid like the genes but in this case RNA (ribose nucleic acid). They are known to be necessary intermediates in the transfer of the protein manufacture instructions from the genes to the cell.

Apart from this important function the endoplasmic reticulum acts as a communication network for the transport of substances within the cell.

(d) *Mitochondria* are oval organelles with large internal surfaces. They are present in all living cells but particularly those that are very active, and their function is aerobic respiration.

On the internal surfaces of the mitochondria are the enzymes involved in the down-grading of sugar and the release of energy for all the other cell functions.

(e) *The cell membrane* surrounds the rest of the cytoplasm and in some places may be continuous with the endoplasmic reticulum. It has a highly organised structure of protein and fat molecules with numerous pores. These allow the passage of water but most other substances that pass across the cell membrane do so actively; that is, their entry is controlled and not necessarily in the same direction as the diffusion gradient.

14. The plant cell. Besides the features listed above the plant cell tends to have a number of other characters associated with the nature of plants and their method of nutrition (*see* Fig. 3). These are as follows:

(a) *Chloroplasts* are bodies similar in size to mitochondria that are found in the surface regions of leaves and shoots. They are packed with the green pigment chlorophyll which is essential in photosynthesis.

(b) *Vacuoles* are spaces full of cell sap, which is a dilute solution of salts and other soluble substances. Animal cells may have a number of small vacuoles but those of mature plant cells are very large and take up most of the volume of

the whole cell. This is possible owing to the rigid supportin
cell wall which allows the thin streams of cytoplasm an
vacuoles to lie within them.

(c) *Cell walls* must not be confused with cell membranes a
the former are secretions of cellulose (a non-living materia]
and other substances. The cell wall lies outside the mem
brane and is readily permeable to the passage of water
sugar and even larger molecules. The cellulose is fairly rigi

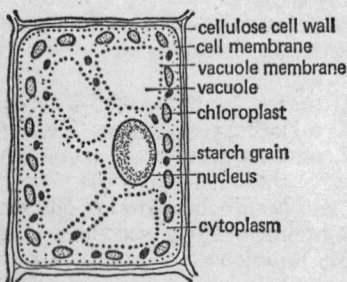

FIG. 3. The plant cell.   FIG. 4. An animal epithelial cell.

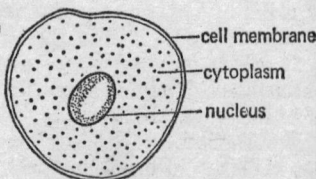

and the whole cell contents are full of water and press hard
against the restraining wall. This results (like a water-filled
rubber bag) in a system of considerable rigidity. In woody
cells the wall is impregnated with a very rigid water-proof
substance called *lignin*. The secretion of this naturally
leads to the death of the cell as it cuts off its supplies of food
and oxygen.

15. The leaf mesophyll cell. We may now go on from a
general consideration of the features of plant cells to consider
an actual example. That chosen will be the mesophyll cell
from a green leaf.

Most broad leaf plants such as the laurel have a definite top
and bottom to their leaves. The top is darker and shinier and
the surface cells are those of the upper epidermis while those at
the bottom are the lower epidermis. These secrete the wax
cuticles but do not photosynthesise, having no chloroplasts.

Below the epidermis are several layers of photosynthetic
cells and it is one of these that should be examined as our

ypical plant cell. The drawing in Fig. 3 is made from a section
f laurel leaf prepared by sectioning with a razor blade and
rawn under the microscope at a magnification of ×200. The
eatures that have been described for a plant cell as well as
ome of the more general features of cells should be made out.
Once again it should be realised that details of ultra-structure
imply cannot be made out at these magnifications.

16. The animal cell. Animal cells have the typical features
outlined in **13**. They do not possess chloroplasts, or have large
vacuoles or cell walls. Animal cells are thus not as rigid as those
of plants.

17. Human epithelial cells. As an example of an easily
obtained and typical animal cell those from the cheek epithelia
of man are useful.

Epithelia are lining cells. Those of the cheek are easily
detached as they are renewed daily. The cells will be made out
as small pale blue squares with a darker blob of the nucleus in
the centre. The cytoplasm is more or less featureless at a magnification of ×200 although it may appear somewhat granular.
A drawing of such a preparation is shown in Fig. 4. The
contrast with, and lack of, plant cell features is quite apparent.

18. Summary chart of plant and animal characteristics.

See table on page 14.

PROGRESS TEST 1

1. What are the major differences between animal and plant
   nutrition? (2)
2. How does the efficiency of anaerobic respiration compare
   with that of aerobic respiration? (3)
3. What is meant by an excretory product? (4)
4. What sorts of animals share the most rapid response to
   stimuli? (5)
5. Why do green plants not need to move in the same way as
   animals? (6)
6. What are the main distinctions between sexual and asexual
   reproduction? (8)
7. What are nucleic acids? (8)
8. Why are viruses hard to classify? (11)

Character of living organism	Plant	Animal
Nutrition	By photosynthesis, chlorophyll present. Carbon dioxide incorporated using light energy.	Food is taken in solid form and broken down. A digestive system is present. A complex diet is necessary.
Respiration	Mostly aerobic (with oxygen) but at a low rate and can withstand periods of anaerobic respiration.	A higher respiration rate than plants. Mostly aerobic.
Excretion	Main excretion is oxygen from photosynthesis. The carbon dioxide from respiration is utilised in light. No nitrogen or salt excretion.	Carbon dioxide and water, also heat from respiration. Excess of salts from the diet and nitrogen-containing products such as urea from internal reactions.
Response	Very limited responses by change in rate of growth. Only responds to light and gravity etc., and very slowly.	Much more elaborate perception of environmental changes, and fast and complex responses of whole animal.
Movement	Very limited growth movement due to differential changes in cell sizes. Necessarily extremely slow.	Extensive movement of a part or the whole of the animal by muscles or other contracting systems. Fast movements normal.
Growth	Plants may continue to grow throughout their lives. The final shape of the plant (such as a tree) depends on surrounding factors.	Animals have limited growth and their mature shapes do not vary in the same way as those of plants.
Reproduction	Both sexual and asexual systems are widespread in plants. Many higher plants are hermaphrodite.	Lower animals may show both forms of reproduction but in all above the level of worms sexual processes are the rule. Hermaphroditism is much less common than in plants.

9. List the major organelle systems of the cell. (13)
10. How do plant and animal cells differ? (14, 16)
11. What are the seven characteristic activities of living things? (18)

EXAMINATION QUESTIONS

1. In what way does a living organism differ from a machine such as a motor car? What do the two systems have in common? (*after O. & C.*)

2. List four structures normally found in living cells and briefly describe their functions. In what ways do plant cells differ from those of animals? (*O. & C.*)

CHAPTER II

THE CHEMISTRY OF LIFE

THE COMPOSITION OF LIVING THINGS

1. Introduction. In this chapter the basic chemistry and properties of the main classes of chemicals involved in metabolism will be described with the way in which reactions can occur at comparatively low temperatures and yet with great velocity by the use of enzymes. These two aspects of biology are fundamental to an understanding of the remainder of the course, and the information and concepts developed in this chapter should be referred to in the study of subsequent chapters.

2. Composition of living matter. The physiological processes described in Chapter 1 (and which are taken as the characteristics of living matter) depend on the interaction of chemicals. It is vital to have some understanding of the chemicals that are involved in metabolism and of the special types of reactions that they undergo.

Until 1840 it was thought there was something "peculiar" about the organic or carbon compounds that we find in living matter because they would not react outside an organism and could not be analysed. This has been shown to be incorrect and nearly all the reactions that occur within cells have been demonstrated in the laboratory outside living material. It has also been possible in recent years to synthesise certain enzymes and to analyse fully very complex molecules such as haemoglobin and nucleic acid. An analysis of two animals and a plant is given below:

	Carbohydrate	Fat	Nucleoproteins + Protein	Salts	Water
Fish (Haddock)	1%	1%	16%	0·5%	81%
Bird (Chicken)	1%	7%	17%	0·6%	74%
Plant (Cabbage)	4%	0%	3%	0·6%	93%

It can be seen that most of a living organism consists of water but that of the organic matter, the three classes carbohydrates, fats and proteins/nucleoproteins, are of the greatest importance.

MAJOR CHEMICAL CLASSES

3. The carbohydrates. These contain the elements carbon (C), hydrogen (H) and oxygen (O), combined in such a way that the ratio of H to O is always 2 : 1 as in water.

(a) *Classification of carbohydrates.* The fundamental unit of the carbohydrate is the single sugar such as glucose or fructose. These both have the basic formula $C_6H_{12}O_6$ although the structural formulae are different. Because they have six carbon atoms, such sugars are called *hexoses*. Another series of sugars is based on a five-carbon skeleton and these are the *pentoses*. As the single sugar unit is called a "saccharide" we can construct a classification of carbohydrates according to the numbers of saccharides they contain. Thus:

(*i*) *Monosaccharides* are single sugars such as glucose and fructose which are both hexoses and the pentose ribulose or ribose.

(*ii*) *Disaccharides* are double sugars which have two units so that if two hexoses are involved they would have twelve carbon atoms. As single sugars condense or join together with the loss of a water molecule we can represent the reaction as follows:

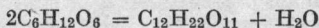

$$2C_6H_{12}O_6 = C_{12}H_{22}O_{11} + H_2O$$

The reverse reaction would be the splitting of a disaccharide with the addition of water to give two molecules of a single sugar. This is called *hydrolysis* and is fundamental in all digestive processes.

Common disaccharides are sucrose or cane sugar which is a combination of glucose and fructose; maltose or malt sugar which is a combination of two glucose units; and lactose or milk sugar which is a combination of glucose with the monosaccharide galactose.

(*iii*) *Polysaccharides* are multiple sugars which may have from ten to many hundreds of units in their molecules. One of the most familiar is starch which is a multiple of twenty or so glucose units, *i.e.* $(C_6H_{10}O_5)_{c \cdot 20}$. Glycogen has a very similar structure.

Larger and more complex is the polysaccharide cellulose of plant cell walls. This has a basic unit of a particular configuration of glucose which gives very long chains and cross linkages. Cellulose is not readily digested by animals and its properties are quite different to starch and glycogen.

(*iv*) *Summary of inter-relationships of carbohydrates.*

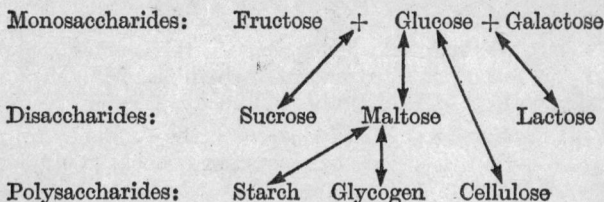

Monosaccharides: Fructose   +   Glucose + Galactose

Disaccharides: Sucrose   Maltose   Lactose

Polysaccharides: Starch   Glycogen   Cellulose

(*b*) *Properties and uses of carbohydrates.* Carbohydrates are perhaps the most common of all classes of organic molecules in living matter and have a wide range of properties and uses in living organisms. These are outlined in the subsections below:

(*i*) *Respiration* (*see* Chapter VI). Glucose is the most common respiratory substance of cells. Enzyme pathways reduce glucose first to $2C_3H_6O_3$ (in the form of pyruvic acid) with the release of some of its contained energy. Then the intermediate product is completely oxidised to carbon dioxide and water with the release of much more energy. Even if other fuels such as fats are being respired, glucose respiration usually takes place at the same time.

(*ii*) *Osmotic effects.* The mono- and di-saccharides exert strong osmotic pressures when separated from water by a semi-permeable membrane. This phenomenon is important in the uptake of water from the soil by plant roots and is generally important in the way in which water enters or leaves a cell (called the *water relationship* of the cell).

(*iii*) *Transport.* The sugars are soluble and this is the form in which carbohydrates are transported by the bloodstream of mammals and the phloem of plants.

(*iv*) *Storage.* The polysaccharides starch (in plants) and glycogen (in animals) are inert molecules used for storage of carbohydrates. These can be mobilised for energy or synthesis as required. Starch tends to be the main food store of seeds, rhizomes and tubers, while fruits and a few plants such as cane and beet store sucrose. Glycogen is stored in the liver and muscles of mammals.

(*v*) *Support.* The polysaccharide cellulose makes up the

cell wall of plants. This is a strong fibrous substance which (on distention of the cell by osmosis) provides a supporting framework. Some plant cells, especially those in the ribs of leaves, have extra supporting cellulose laid down in the corners.

4. The fats. These contain the elements C, H, and O but the H and O are not in the proportions found in water. There is much less O present than in carbohydrate molecules.

(a) *Chemistry of fats.* Fats are a combination of fatty acids such as stearic acid $C_{17}H_{35}COOH$ or palmitic acid $C_{16}H_{31}COOH$ with glycerol, $CH_2OH.CH.OH.CH_2OH$. This combination or condensation and the reverse, or hydrolysis, may be represented as follows:

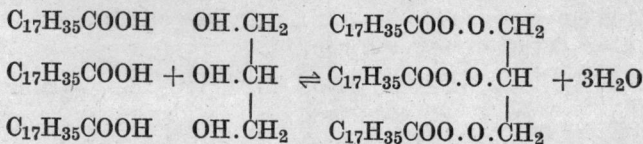

$$
\begin{array}{ccc}
C_{17}H_{35}COOH & OH.CH_2 & C_{17}H_{35}COO.O.CH_2 \\
| & & | \\
C_{17}H_{35}COOH + OH.CH & \rightleftharpoons C_{17}H_{35}COO.O.CH & + 3H_2O \\
| & & | \\
C_{17}H_{35}COOH & OH.CH_2 & C_{17}H_{35}COO.O.CH_2
\end{array}
$$

Three fatty acids do not always have to combine with glycerol, either one or two fatty acids will also do for glycerides or fats.

Oils are closely related to fats and are liquid at normal temperatures whereas waxes are complexes of fatty acids and alcohols and they are very stable.

(b) *Uses of fats.* In living systems these come under several main categories. These are described in the sections below:

(i) *Respiration.* Fats have more than twice as much energy as carbohydrates for the same unit weight. Fats can provide metabolic energy on respiration.

(ii) *Food store.* Because of their high energy content fats are often used as a concentrated food store. In mammals they may be stored under the skin, where they also provide insulation, or around major organs such as the heart or kidneys. In plants, fat is stored in certain seeds, for example cotton and the soya bean.

(iii) *For structural purposes.* Fats are incorporated into all the membrane systems of the cell, and so are included in the membrane, endoplasmic reticulum, mitochondria, nuclear membranes etc.

(iv) The presence of fats in the intestine is important in the assimilation of the fat soluble vitamins A and D.

(v) In the form of waxes fatty substances are involved in *waterproofing* the insect and plant cuticles.

5. The proteins. These are the most complex of the organic molecules of living things and may have molecular weight in excess of 10^6. They all contain the elements C, H and O as well as N (nitrogen) and may incorporate other elements such as S (sulphur) or P (phosphorus).

(a) *Chemistry of protein.* Proteins are formed by the condensation of amino acids whose basic formula is $NH_2.R.COOH$ where R can be a number of different radicles. The condensation takes place by formation of peptide linkages with the elimination of water while hydrolysis of proteins takes place in the reverse direction

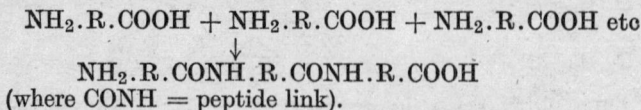

$$NH_2.R.COOH + NH_2.R.COOH + NH_2.R.COOH \text{ etc}$$
$$\downarrow$$
$$NH_2.R.CONH.R.CONH.R.COOH$$

(where CONH = peptide link).

There are some twenty or so different amino acids and as these can be linked together in all sorts of ways, and as hundreds or thousands may be involved in the structure of a single protein, there are an almost infinite number of possible proteins that can be formed.

Individual amino acids vary considerably in their structure and some are complex involving aromatic (cyclic) groups and incorporating sulphur. On the whole the body of a man can make the simpler amino acids such as glycine but not the complex ones like methionine. Plants on the other hand can make all types of amino acids from very basic constituents.

(b) *Properties and uses of proteins.* Proteins are the most important of all biological chemicals and have an enormous number of different properties and uses. A few of these to illustrate their diversity are listed below:

(i) *Structural and contractile.* Keratin of wool and collagen of tendon and connective tissue are structural proteins while the actin/myosin proteins of muscle are contractile.

(ii) *Respiratory pigments*, such as myoglobin in muscle

and haemoglobin in the blood, are proteins which increase the oxygen-carrying capacity of these tissues.

(*iii*) *Nutritional or storage proteins*, such as casein in milk and albumen in eggs, provide a store of amino acids.

(*iv*) *Antibodies* or protective proteins are present in the blood of mammals and in one form or another in all living things. These proteins "recognise" and inactivate foreign proteins, thus protecting the organism from parasitic infections.

(*v*) *Enzymes*. Proteins are incorporated into the thousands of enzymes which allow the reactions of an organism's metabolism to take place. All enzymes have a substantial protein part, or are entirely protein structures.

6. Nucleoproteins. These are a very special type of protein which include a nucleic acid and are best considered as a separate class of chemicals. They are involved in transmission of heredity factors and the manufacture of enzymes and other proteins within cells, so they are of the greatest importance.

(*a*) *Chemistry of nucleoproteins.* The protein part of the molecule is attached to long chains of bases which incorporate chemical information. The bases are arranged in meaningful sequences each base being attached at one end to a ribose sugar and paired with another base. The sugars are joined to each other via phosphates.

The four bases are called *nucleotides* and thymine is paired with adenine and guanine with cytosine. With these particular bases and one form of ribose sugar we have the DNA (deoxyribose nucleic acid) of the nucleus. Other forms of nucleic acid result from different ribose sugar, and uracil replaces thymine. These are the RNAs (ribose nucleic acids) of the cytoplasm.

(*b*) *Properties of nucleoproteins.* Nucleic acids are enormous molecules with molecular weights in the $100,000 \times 10^6$ class. One "molecule" may contain many items of genetic or chemical information. In general the DNA code of the nucleus is inherited directly from the parents and is transmitted to the RNA of the cytoplasm. Here it is interpreted on the ribosomes of the endoplasmic reticulum and the complex sequences of amino acids are built up into proteins of a given organism or cell.

Another essential property of DNA is its ability to replicate itself exactly from bases, phosphates and sugars present in the nucleus. This happens every time a cell divides by mitosis or meiosis (*see* Chapter X).

Although the molecules of the nucleic acids are very stable they may be changed by radiation or certain chemicals. This, of course, changes the chemical coding they carry, and tends to produce a nonsense sequence. Such changes are called *mutations* and are usually harmful. A very few tend to be beneficial and the evolution of life has depended on this fact.

SYNTHESIS OF PROTEIN IN CELLS

7. Synthesis of protein

(*a*) *Transfer of information from nucleus to cytoplasm.* Here we shall consider in more detail the relationship between the nucleus and the cytoplasm and how the cell transmits information from the former to the latter in order to manufacture the correct proteins.

Within the nucleus is the DNA with its information in the form of a series of base pairs—adenine with thymine, and guanine with cytosine. The information is passed to messenger RNA. Messenger RNA is formed from ribose sugars, phosphate and the bases uracil and adenine and guanine and cytosine.

At the start of this RNA formation the helical molecule of the DNA splits and unwinds by action of the enzyme RNA polymerase. Half the strand is not used but the other half, which has the information sequence, acts as a template upon which the RNA is built up, uracil coming next to adenine and cytosine to guanine. This happens as in Fig. 5.

(*b*) *Synthesis within the cytoplasm.* The messenger RNA comes away from its DNA template and passes out into the cytoplasm where it becomes attached to a ribosome. This process is repeated many times so that the cytoplasm contains very many ribosomes which carry this particular section of the gene code.

Amino acids, the building blocks of protein, are taken in at the cell membrane or, in some cases, made within the cell itself. (There will be approximately twenty of these amino acids.)

At this stage another form of RNA is important; this is

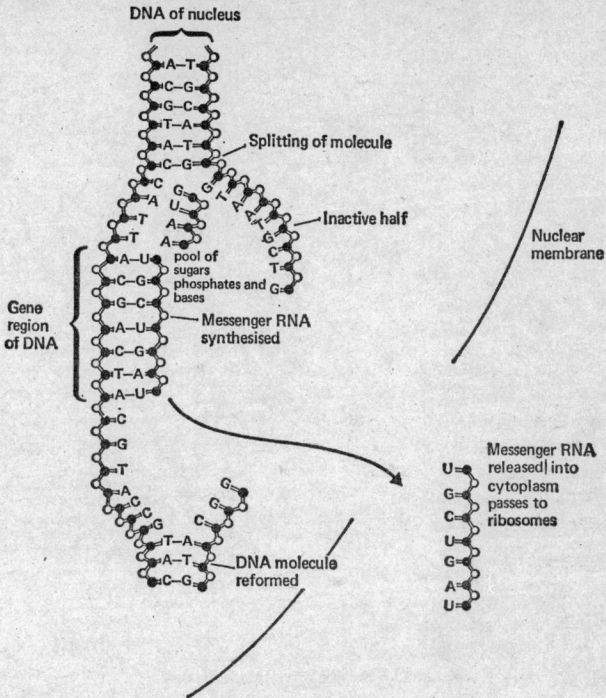

FIG. 5. DNA replication in the nucleus.

called "transfer RNA" to distinguish it from the messenger RNA whose role we have considered. Transfer RNA is coded at one end to fit into the appropriate site on the ribosome while the whole transfer molecule is specific for a single amino acid.

The transfer RNA picks up its particular amino acid at the cell surface and carries it to a ribosome with the correct code position. The code has been found to consist of three bases and is called the triplet code. Using the same piece of messenger RNA synthesised in the previous example, we can carry the process further to actual manufacture of part of a protein (see Fig. 6).

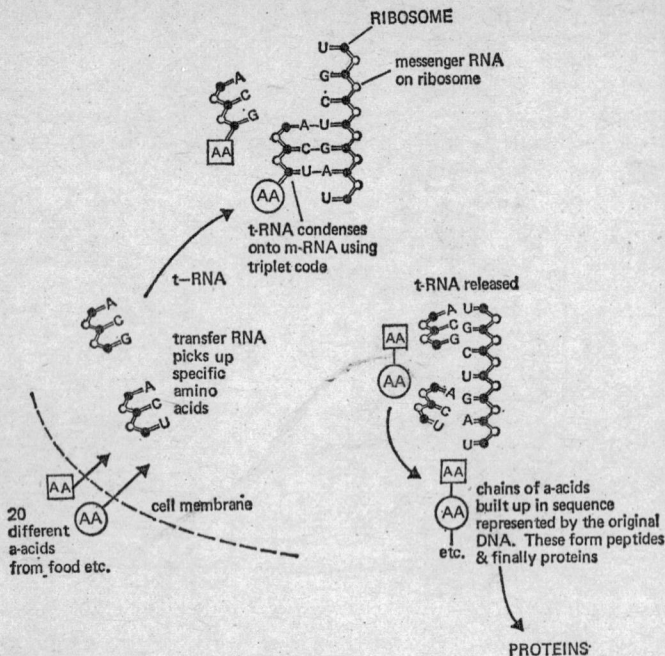

FIG. 6. Protein synthesis in cells.

THE NATURE AND FUNCTIONING OF ENZYMES

8. The chemistry of enzymes. Enzymes always contain a protein part to their molecule and may be entirely protein. Many enzymes contain, or work in connection with, a subsidiary factor called a *co-enzyme*. Vitamins of the B group are co-enzymes to many respiratory enzymes which are concerned with the transfer of energy. Sometimes metals such as iron or copper are active as co-enzymes.

The important feature of all enzymes is that they act as catalysts which allow reactions to proceed by lowering the energy required for the particular reaction to occur. Thus at body temperature all sorts of chemical reactions occur at speeds which would be quite impossible without the presence of enzymes.

9. Properties of enzyme controlled reactions. There are a number of properties of enzyme reactions which it is important to understand as they relate directly to the conditions necessary for all physiological reactions to take place. Thus:

(a) *Specificity.* Enzyme molecules exactly fit the shape of the molecules (or substrates) which they cause to react. For this reason enzymes are specific and each will bring about only one type of reaction. Maltase will hydrolyse maltose but has no effect on sucrose or lactose.

(b) *Temperature dependence.* Enzymes operate most efficiently at optimum temperature. For mammals this would be around 40°C which is the temperature of their bodies, and for plants the optimum is much less. Below about 5°C and above 50°C enzymes do not work effectively; in the first instance their rates are extremely slow while at high temperatures they are denatured or destroyed (Fig. 7).

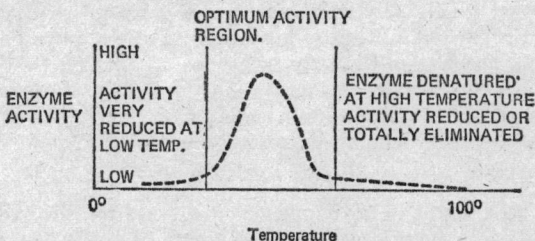

FIG. 7. Effect of temperature on enzyme activity.

(c) *Poisoning by chemicals.* Besides high temperatures many enzymes are denatured by heavy metals (such as lead). Cytochrome, a key respiratory enzyme, is poisoned by cyanide. The toxicity of substances such as arsenic and strychnine to man or DDT to insects is due to enzyme denaturisation.

(d) *pH.* Enzymes have an optimum pH which is the degree of acidity or alkalinity at which they operate best. This is often around the point of neutrality but sometimes, as in the case of pepsin, an extreme pH is required.

(e) *The amount of enzyme* to bring about a given reaction may be extremely minute. Most enzyme molecules turn over hundreds and hundreds of substrate molecules each minute, and they are renewed themselves during the same time.

(f) *Hydration.* Enzyme reactions are surface effects. In order to be effective, molecules must have large areas of contact such as occurs in solution. Enzymes cease to work when dehydration of protoplasm takes place.

10. Some types of enzymes described. All types of reactions in living things are brought about by enzymes. Approximately 1,500 individual enzymes have been isolated and recognised at the present time but it is estimated that there are at the very least some hundred times more than this that remain to be revealed. Synthesis, muscle contraction, energy release, active transport across membranes, gas transport in blood, clotting, sugar uptake by cells and every other metabolic process conceivable take place due to the activities of specific enzymes.

A very common and familiar class of enzymes are those which bring about hydrolysis during digestion (*see* Chapter V). Less familiar but quite as important are the respiratory enzymes that pass hydrogen from one substrate to another (*see* Chapter VI) with the release of energy. Other classes of enzymes are important in making amino acids of one type into others, or adding a specific chemical group or bringing about molecular rearrangements (*e.g.* glucose into fructose).

11. Enzymes and the conditions necessary for life. An understanding of the properties of enzymes described above will help to explain why life is only possible under certain very specific conditions. Thus for the human body to function normally it must be at the correct temperature, pH and degree of hydration, and no toxic substances must be present. Mammals are exceptionally complex in the exactness of conditions they require to remain alive and even micro-organisms (which cause decay and spoilage of stored food) may be destroyed by a gross upset of the conditions under which their enzymes will operate. These micro-organisms can be guarded against by:

(a) *Dehydration,* as for biscuits, flour or chocolate which prevents the survival and multiplication of bacteria and fungi.

(b) *Temperature changes*; either freezing which arrests activity or heating which destroys micro-organisms through denaturisation of their protein and enzymes will preserve food.

(c) *Introduction of toxins* (but not those toxic to us!) such as vinegar, which will prevent the multiplication of micro-organisms.

PROGRESS TEST 2

1. List the main constituents of living matter. (2)
2. How do monosaccharides differ from polysaccharides? (3)
3. What are the primary uses of carbohydrates by living organisms? (3)
4. What are the products of the hydrolysis of fats? (4)
5. What are the primary uses of fats by living organisms? (4)
6. What are the units of which protein is made up? (5)
7. What is the importance of nucleoproteins? (6)
8. What are the stages of protein synthesis in the cell? (7)
9. What are enzymes? (8)
10. What are the main properties of enzymes? (9–11)

EXAMINATION QUESTIONS

1. Of what importance are mineral salts to living organisms? Devise an experiment to show the importance of a named mineral salt in the growth of a plant. (*A.E.B.*, 1970)

2. The graph below shows the effect of temperature on the rate of enzyme reaction—the enzyme being derived from a mammal.

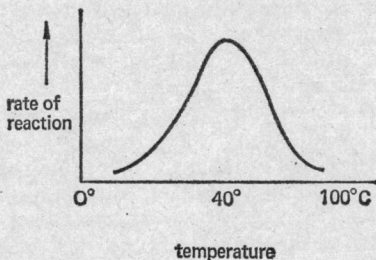

Discuss the interpretation and significance of this graph. (*after Cambridge*)

CHAPTER III

THE NUTRITION OF THE GREEN PLANT, PART 1

RAW MATERIALS

1. The importance of green plants to other organisms. Green plants in the form of trees, herbs and grasses cover much of the earth's land surface. In the surface waters of lakes, rivers and of the seas they exist in the form of microscopic floating algae whose weight, per unit area, is equal to that of plants on good soil. Fringing the land masses much larger algae grow as seaweeds, extending from the splash zone above the high tides to many feet beyond the lowest tide mark.

All these types of plant, on land and in water, feed by the common method of photosynthesis, a process upon which all forms of life on this planet ultimately depend.

(*a*) *Photosynthesis and food production.* In essence photosynthesis consists of the combination of carbon dioxide with the hydrogen atoms of water to form a number of organic molecules such as sugar and starch. These latter may in turn be combined with various inorganic ions to produce amino acids and eventually proteins. Molecular rearrangement allows fat formation. Thus from simple substances present in the atmosphere, the soil or the water, in which the plant may be growing, it can build up all the types of molecules required by living organisms.

On both land and water green plants are the basic members of food chains and webs, being eaten by herbivores which in their turn provide nutrition for carnivores. The dead bodies of plants, and of the animals which have fed on plants, in their turn provide food for the saprophytes that live in the soil or on the sea bed.

Green plants have much more extensive powers of chemical synthesis than animals; for example plants do not require to be supplied with vitamins, they make them within their

Land plants	Nutrient	Source
Trees, herbs, grasses, ferns, mosses etc.	Carbon dioxide	The atmosphere, into which it is returned by decay of dead and waste matter and by plant and animal respiration. Atmospheric CO_2 is also increased by man's industrial activities and the amount present appears to be steadily rising.
	Water	The soil is the source of the plants' water, and it retains various amounts according to its particle size and other factors. Land plants can only take up water via their roots.
	Nitrogen as NO_3^- Phosphorous (PO_4^{---}) Sulphur (SO_4^{--}) Calcium (Ca^{++}) Potassium (K^+) Magnesium (Mg^{++}) Iron (Fe^{+++}) and other ions	All these ions are held in solution in the soil water that surrounds the particles. The sources of these ions are from decay of organic matter by saprophytic micro-organisms. Man adds extra quantities of the mineral ions as fertilisers to promote optimum growth of crops.
Water plants		
Seaweeds and minute algae (phytoplankton). Higher plants found in lakes and rivers.	Carbon dioxide	CO_2 is relatively soluble in water (100 parts gas per 100 ml water). The carbon dioxide content of water is in equilibrium with that of the atmosphere. CO_2 is returned to water by decay on the sea bed or its freshwater equivalent and by the respiration of aquatic organisms.
	Water	From the medium itself.
	All mineral ions	These are returned to water by decay of organic matter and by the run-off of nutrient salts from the land. There is a very slow solution of the solid minerals of rocks. In all, the sea contains 3·5% of mineral salts in solution and freshwater has almost negligible quantities.

cells. Animals tend to need a varying number of vitamins in their diet simply because they are quite unable to synthesise them.

(b) *Photosynthesis and the atmosphere.* Besides this provision of food for all other types of living things, the plants, by removal of the hydrogen from water molecules, cause the release of oxygen into the atmosphere. The 20 per cent oxygen content of the earth's atmosphere at sea level is entirely due to the photosynthetic activities of green plants and it enables both plants and other organisms to respire aerobically. Life, as we know it, would not be possible without this oxygen for respiration (*see* the carbon cycle in Fig. 27, page 98).

Plants and animals and saprophytes all give out carbon dioxide as a result of respiration. The atmosphere contains some 0·04 per cent of this gas and each year 9×10^{11} tonnes are "fixed" into organic form by the process of photosynthesis.

2. The sources of the plant's nutrients. The major nutrients required by plants are water and carbon dioxide as well as the elements nitrogen, phosphorus and sulphur. These latter are taken in in the form of diffusable ions, *i.e.* NO_3^-, PO_4^{---} and SO_4^{--} and are of particular importance in protein manufacture. There are many other inorganic ions which are used to a lesser extent by green plants. These nutrients and their sources are summarised in the table on page 29.

THE PLANT AND THE SOIL

3. The soil and the plant root. A fertile loam soil contains many insoluble mineral particles which may vary from less than 0·002 mm to over 2 mm in size. The properties of a given soil depend to a large extent on the size of the particles that it contains because this determines how much air and water will be retained and these directly affect the growth of plants.

(a) *Direct effects of particle size on the root.* If the particles are very small, as in clay, the soil will retain a great deal of water but very little oxygen. The soil will be heavy and waterlogged and plant roots will not be able to respire because of the lack of air spaces from which they can draw oxygen.

As we shall see below roots have a high rate of respiration because of their functions, so plants in waterlogged conditions will not thrive due to poor root development.

Soils with very large particles, such as sands, have plenty of air but very little water. As the plant depends on a continuous supply of water from the soil it will once again fail to grow properly. The ideal soil with the right balance of water and air is loam and this has a mixture of particles of all sizes.

(b) *Indirect effects of particle size on the root.* A fertile soil with its intermediate composition encourages the growth of plants and the active processes of decay provide valuable humus for the soil. Humus is a brown organic complex which contains all sorts of ions that are needed for plant growth. The more humus a soil contains the richer and more fertile it will be.

The process of decay is brought about by the great number of saprophytic organisms in the soil which liberate the chemicals from the hosts on which they feed and which also by their own deaths contribute to humus formation.

(c) *Root hairs and the soil.* At the tip of the root there is a protective cap which lubricates its passage during growth through the soil. Behind this is the meristem.

(i) *The root meristem* is a region of active cell division. Here the cells are small and undifferentiated, but further up the root the various individual cell types and tissue systems are clearly defined.

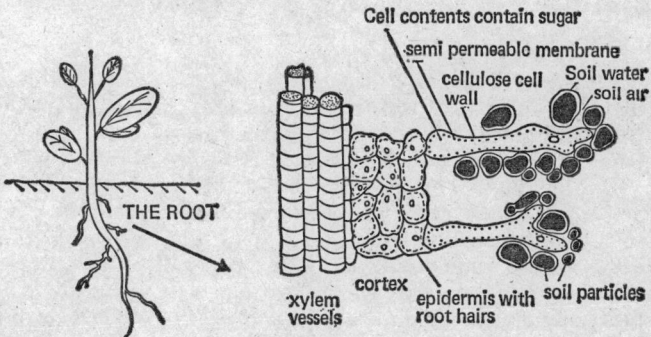

Fig. 8. Relation of soil water to the root hairs.

(*ii*) *Root hairs.* Some distance behind the meristem the epidermal cells become elongated into root hairs. These are elongated cells with thin walls which make a very intimate connection with the particles of the soil. This relationship between root hair and soil is shown in Fig. 8. The root hairs greatly increase the surface area of this part of its structure and are freely permeable to the entry of water. Their cell contents are full of sugar brought down from the overground parts of the plant and thus osmosis occurs across the semi-permeable cell membranes.

Osmosis is the diffusion of water across a semi-permeable membrane (one permeable to water) from a weak to a stronger solution of a chemical whose own molecules are too large to pass across the membrane. Thus sugar within a semi-permeable membrane such as a pig's bladder will draw water across and the same phenomenon is seen when sultanas (dried grapes) or prunes (dried plums) are soaked in water. The swelling of these fruits in water is due to osmotic entry of water. Osmotic pressure is further dealt with in Chapter VII.

(*iii*) *Soil water.* In loam soil all the particles will be surrounded with a thin layer of capillary water and it is from this source that the water enters the root hair cells. The osmotic pull of the root hairs can remove most of the capillary water but there comes a point of balance. At this stage the soil will appear very dry and the plant is likely to wilt.

(*iv*) *Transplanting seedlings.* When a gardener transplants seedlings he is careful to retain undisturbed soil around their roots so that the delicate hairs are not torn off. The seedling is watered-in to its new position to ensure a plentiful supply and nowadays it is common practice to germinate many plants in fibrous pots which can be moved *in toto* to their final position.

(*d*) *Mineral ions and their entry into roots.* Mineral ions dissolved in the soil water may pass into the root hairs by passive diffusion. This would be the case of ions that are being removed from the roots to other parts of the plant.

In other cases, for example with the K ions, a process of active transport occurs where ions are pumped into the root cells by a process that involves respiratory energy; *i.e.* the cell must use some of its energy reserve to allow the process to occur. This is one of the reasons why efficient root functioning depends on a plentiful supply of oxygen in the soil.

(*e*) *Deficiency disease.* Where certain elements (in their ionic form) are lacking in a particular soil the plants that grow there will show various mineral deficiency diseases. These have symptoms connected with the role of the particular element within the plant. Most common are the following. Figure 9 shows typical effects of mineral deficiencies.

Deficiency	Symptom	Reason
N, P, S	Poor growth	These minerals are involved in protein formation.
Fe, Mg	Yellow leaves	These minerals are necessary for chlorophyll formation.
Ca	Stunted plants	Calcium is involved in cell wall structure. Plants without proper cell walls tend to collapse.
K	Abnormal shapes, poor root growth, no flowers	Potassium is involved in a number of ways. Some of the symptoms are due to failure of the phloem transport system in the plant.

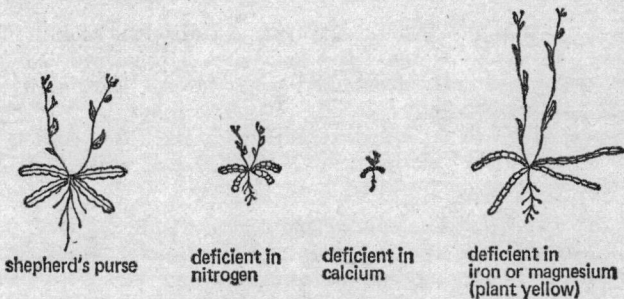

shepherd's purse   deficient in nitrogen   deficient in calcium   deficient in iron or magnesium (plant yellow)

FIG. 9. Deficiency of mineral causes abnormal growth in plants. Normal growth is shown on the left.

Deficiency diseases, and there are many more than are shown above, can be diagnosed and treated by the addition of the appropriate mineral salt to the soil. The return of elements into the soil is covered in Chapter VIII.

PLANT ANATOMY RELATED TO FUNCTION

4. Transport systems within the green plant. The gross features of organisation of green plants have already been described in Chapter I. Most of the metabolic activities of the plant involve transport of raw materials, nutrients, plant hormones and other substances from one place to another.

Thus water and ions travel from the roots to the leaves, and from the latter pass the synthesised sugars, amino acids and fats to storage organs and growing tissues. The growing regions produce auxins (plant hormones) which co-ordinate the growth patterns of the plant and must be transported to the regions they affect. Flowers are supplied with sugar for nectar secretion and later the ripening fruits and maturing seeds will also require all types of food substance. In the spring, storage organs such as seeds, tubers, bulbs etc. will in turn send out new growth and foods will be conducted from the storage organ out to this new tissue.

In all a very large percentage of the tissues of the root, stem and leaves of a typical higher plant are composed of conduction tissue.

5. Root anatomy, general features. A transverse section of a young root in the region of the root hairs shows a wide cortex of unspecialised cells often filled with starch. The details of root anatomy are shown in Fig. 10. The inner layer of this cortex is called the *endodermis* and this has a waterproof substance called *cork* laid down in parts of its cell walls allowing control of water movement.

(a) *The stele.* The conducting region within the cortex is termed the *stele* and it consists of the central star-shaped xylem and outer separate areas of phloem.

(b) *The xylem* consists mainly of hollow elongated vessels whose diameters may be up to 0·5 mm. The walls of these cells are lignified (that is, strengthened with wood) so that xylem vessels provide a means of water transport as well as support for the plant. The latter function may be increased by the presence of fibres among the vessels. Besides such dead cells xylem may also contain living parenchyma cells important in storage.

(c) *The phloem*, which is used for the conduction of sugar and other organic molecules, consists of sieve tubes which do

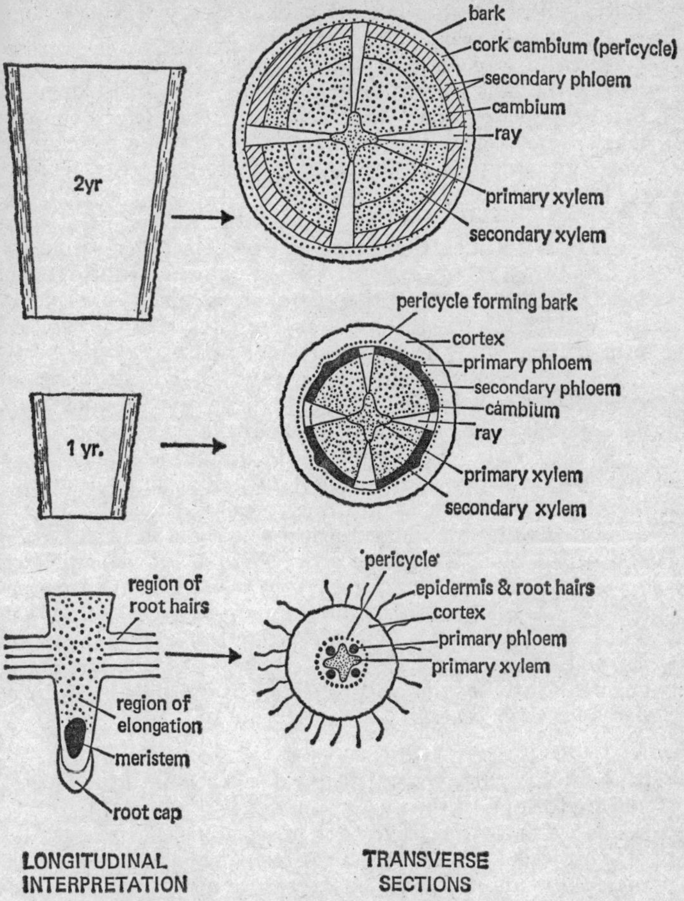

bark
cork cambium (pericycle)
secondary phloem
cambium
ray

primary xylem
secondary xylem

2yr

pericycle forming bark
cortex
primary phloem
secondary phloem
cambium
ray
primary xylem
secondary xylem

1 yr.

region of root hairs

pericycle
epidermis & root hairs
cortex
primary phloem
primary xylem

region of elongation
meristem
root cap

LONGITUDINAL INTERPRETATION

TRANSVERSE SECTIONS

FIG. 10. Anatomy of the root.

the actual conduction, and associated companion cells. The function of the companion cells is not clear but they do seem to assist the working of the sieve tubes. Phloem, unlike xylem, is alive in its functional state. The sieve tubes are elongated cells whose end walls are perforated, and

through these sieve plates strands of cytoplasm pass from
one cell to another.

(d) *Secondary thickening in the root.* As the root grows
older a process called *secondary thickening* starts whereby
more supporting and conducting tissue is added to the primary structure. A protective layer of bark also forms from
cells at the outside of the stele so that the cortex gets cut
off from below and dies.

Secondary thickening starts with the formation of cambium cells between the xylem and phloem. These differentiate between xylem on the inside and phloem on the outside.
Because of this ability the cambium is called a *secondary
meristem.* Its cells have much in common with those of the
primary meristems at root and shoot tips.

During secondary thickening, rays or radiating lines of
parenchymatous cells are also laid down and the root has a
considerable volume of such tissue available for storage.

(e) *The mature root* is thus protected on its outside from
mechanical injury and the entry of predators by the thick
bark. Inside there is a strong xylem core providing a
system of water and ion transport as well as support for the
rest of the plant above the soil. There is sufficient phloem
around the xylem for downwards transport of food to storage organs and growing points. Parenchyma cells in xylem
and phloem can be used for storage, and finally, the seasonal
activity of the cork and the deeper layers of cambium cells
ensures that the root will continue to grow to meet fresh
demands each year.

6. Stem anatomy, general features. The main functions of
stems are to support the overground parts of the plants and to
conduct water to the leaves and to remove the products of
photosynthesis. In transverse section the young stem is
bounded by an impermeable layer of epidermal cells surrounding a cortex. Details of stem anatomy are shown in
Fig. 11. The vascular tissue lies within this cortex.

(a) *The stele.* Unlike the root the vascular tissue is made
up of a number of groups of xylem and phloem lying on the
same radii. These are called *vascular bundles.* The detailed
structure of the xylem and phloem are identical to that
already described for the root.

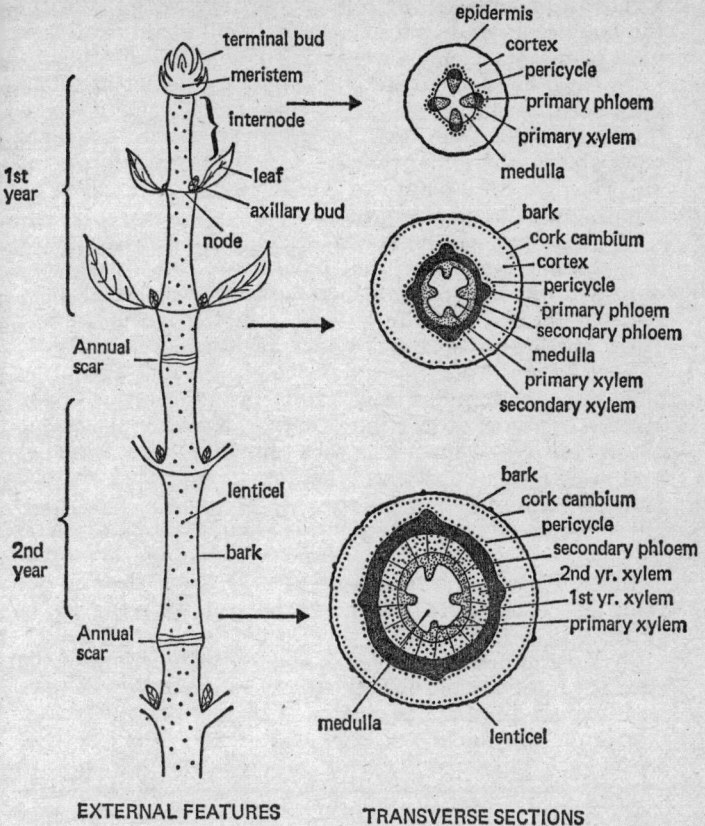

terminal bud
meristem
internode
leaf
axillary bud
node

epidermis
cortex
pericycle
primary phloem
primary xylem
medulla

1st year

Annual scar

bark
cork cambium
cortex
pericycle
primary phloem
secondary phloem
medulla
primary xylem
secondary xylem

2nd year

lenticel
bark

Annual scar

bark
cork cambium
pericycle
secondary phloem
2nd yr. xylem
1st yr. xylem
primary xylem

medulla
lenticel

EXTERNAL FEATURES   TRANSVERSE SECTIONS

FIG. 11. Anatomy of the stem.

(b) *The medulla.* The centre of the young stem consists of parenchyma cells termed the *medulla*. These may serve a storage function.

(c) *Secondary thickening in the stem.* The cambium layer which initiates the process of secondary thickening develops between the xylem and phloem, at first within the vascular bundles but later joining one bundle to another.

On the inside of the cambium new xylem is differentiated

and on the outside new phloem appears increasing the size of the bundles as well as forming new bundles and finally, in most plants, producing a whole ring of new vascular tissue.

This will be the final state in annual plants but in perennials such as shrubs and trees this cambial activity continues throughout the year. In spring the vessels formed are large while in the winter they are very small or not formed at all. The age of the plant can thus be readily determined by counting these annual rings.

(d) *Medullary ray formation.* Extensions of parenchyma cells through the xylem and phloem are called *medullary rays.* The number of rays is increased each year though only the first formed actually extend from the medulla itself. These rays are important as storage regions in woody plants.

(e) *Bark and lenticel formation.* Unlike the root, formation of cork cambium takes place within the outer layers of the cortex. It occurs during the first year in woody plants but not in herbaceous ones. The cork cambium makes layers of corky cells on the outside and these are impregnated with the fatty suberin forming a waterproof, airtight and mechanically protective layer. At intervals around the stem, where stomata were present in the primary structure, loose collections of cork cells form lenticels. Through these, gases from the atmosphere can be exchanged with the living tissues within the stem.

(f) *External features of stems.* One of the functions of the stem is to provide for new growth and the origin of flowers and leaves. Features associated with these activities can be seen on the outside of this plant organ. The length of the stem is interrupted by nodes and here the leaves are formed. In the axes of these leaves buds develop which are consequently called the *axillary buds,* and these may give rise to further complete shoots or to flowers.

At the end of the stem is the apical meristem marked by a terminal bud or inflorescence. This apical region dominates the whole development of the stem region behind it, producing the plant hormones which control growth.

(g) *The mature stem* is thus a more complex organ than the root although it shares functions of transport, storage, support and the ability to increase in size. The stem, however, is able to produce whole new organ systems, leaves, flowers, and if necessary, roots as well. From a stem cutting a whole new plant can be grown. There is good reason for thinking of the stem as the basic plant organ and the roots, leaves and other parts as extreme modifications of this basic organ.

7. Leaf anatomy. The leaf is an organ modified for the control of water loss and for the whole process of photosynthesis. Leaves grow from the nodes of stems and the arrangement of leaves or of leaf pairs gives a definite pattern and the minimum of overshading of one leaf by another. The stalk joining the leaf to the stem is called the *petiole*.

(a) *The petiole* mimics the structure of the stem with a symmetrical arrangement of vascular tissue. Shortly after its exit from the stem the petiole is transversed by a line of weakness, or abscission layer. This is the point where the leaf will part from the parent plant when it is shed and the presence of a layer of cork across most of the petiole prevent water loss and invasion of disease.

By twisting of the petiole the leaf tends to be brought at right angles to the incident light, the best inclination for photosynthesis.

(b) *Lamina.* At its far end the petiole flattens out into the blade or lamina of the leaf, the vascular tissue and supporting elements continuing up this as the midrib. From the latter, many veins are given off and these branch into so many smaller vessels that eventually no cell in the leaf is far from the end of a vein. As these veins are the ends of the vascular system they both supply and carry away some of the raw materials and products of photosynthesis.

If a leaf, such as laurel, is examined, a thick shiny cuticle will be found on the upper side. The lower surface has a more mat-like surface and a thinner cuticle. Leaves of this sort are termed *dorso-ventral*. Grasses, rushes, reeds, irises etc., whose leaves tend to grow directly upwards, have no distinction between the surfaces and for this reason are termed *iso-lateral*.

(c) *Micro-structure of the leaf.* A transverse section through a leaf shows that the cuticle is secreted by the regular layer of epidermal cells. These cells contain no chloroplasts in sharp contrast to the tightly packed elongated layers of palisade tissue that lie below them. Figure 12 shows a leaf section.

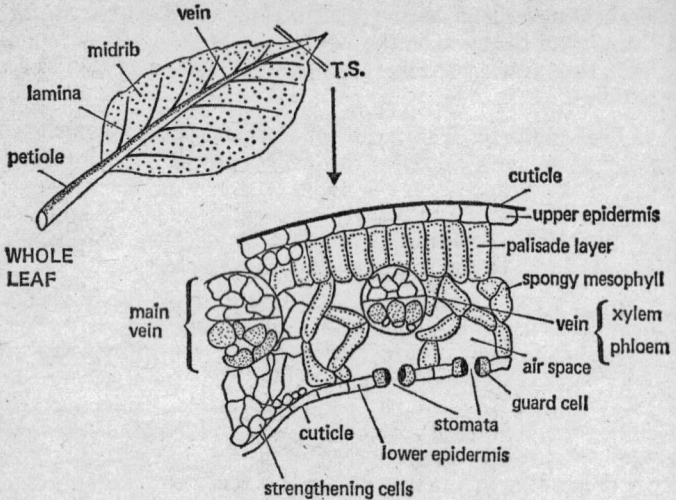

FIG. 12. Leaf section across midrib region.

Below the palisade is the spongy mesophyll layer(s): cells similar in structure to the palisade, but with many air spaces between them to allow the rapid diffusion of gases. Vein endings are between palisade and mesophyll cells.

The lowest tissue of the typical leaf is another layer of epidermal cells and although they do secrete a cuticle they differ from the upper epidermis by the presence of guard cells and stomata.

(d) *Stomata*, with many hundreds to the cubic milli-metre, are holes through which gases can diffuse into and out of the leaf. These holes are able to open or close according to the conditions in which the plant is situated. They will tend to be open in daylight allowing the entry of carbon dioxide and escape of oxygen and water. At night the stomata are usually shut to prevent water escaping.

Surrounding each stoma, and forming the basis of its function, are two guard cells. These are the only types of cells in the epidermis that have chloroplasts and can photo-synthesise. In light they make sugar and become turgid so that they swell, and, due to their peculiar thickening, draw apart from each other. During darkness the sugar is lost or converted to starch so that the guard cells lose water to their neighbours and resume a normal size. The edges of the cells come together, shutting the stoma.

(e) *Summary of leaf adaptations.* As has been seen the leaf is a highly advanced and modified structure. Its major functions are photosynthesis and water control, and its structure and function may be summarised by the following table:

Structure	*Function*
Petiole	Support of leaf, transport, orientation, abscission
Lamina (gross features)	Thin, allows passage of light. Broad, maximum surface for light.
(micro-structure)	Cuticle prevents water loss. Veins allow transport. Palisade and mesophyll adapted for photosynthesis. Air spaces and stomata allow gas diffusion for efficient movement of CO_2 and O_2. Stomatal mechanism controls water loss in conditions not suitable for photosynthesis (*i.e.* darkness, wilting)

PROGRESS TEST 3

1. How does oxygen get into the atmosphere? (1)
2. What are the major nutrient requirements of green plants? (2)
3. What effect does particle size of soil have on root survival? (3)
4. What is a meristem? (3)
5. What effect on the plant does a deficiency of the following elements produce? Ca, Mg, K. (3)
6. In what order would the tissues in a cross section of a two-year-old woody root be found? (5)
7. What is a medullary ray? (6)
8. How does the petiole of a green leaf help bring the lamina into optimum light intensity? (7)
9. What is the importance of stomata? (7)
10. List the adaptations of a leaf for efficient photosynthesis. (7)

EXAMINATION QUESTIONS

1. What are the raw materials and environmental factors necessary for photosynthesis to occur? Describe how the structure of the leaf facilitates the supply of the materials needed for photosynthesis. State briefly what happens to the products of photosynthesis. (*O. & C.*, 1972) (*See* also Chapter IV)

2. In what form does a green plant obtain each of the following: carbon, hydrogen, oxygen, nitrogen, sulphur, phosphorus, potassium, magnesium? Describe the components of a fertile soil. (*after O. & C.*, 1969)

3. Describe briefly the processes by which water enters the root hairs and escapes from the leaves at the top of a tall tree. Your answer should include the following stages:

    (*i*) entry of water into the root hair;

    (*ii*) passage of water from root hair to the xylem;

    (*iii*) passage up the xylem;

    (*iv*) passage from the xylem to the mesophyll cells of the leaf;

    (*v*) passage from the mesophyll cells to the atmosphere through the intercellular spaces to the stomata.

(Cambridge, 1970)

4. Describe the changes that take place in a young stem as it enters the second year of life. (*after O. & C.*)

CHAPTER IV

THE NUTRITION OF THE GREEN PLANT, PART 2

TRANSPIRATION

1. Introduction. In the previous chapter the entry of water and mineral nutrients into the plant was described together with the basic anatomy of root, stem and leaf. In order to understand how the processes of transpiration, photosynthesis and translocation, all a part of the whole nutrition of the plant, are carried out, the contents of the previous chapter must be constantly borne in mind. Appropriate cross references are included.

2. Transpiration is the movement of water through the plant from the soil, via the root and stem and out through the leaves to the atmosphere. Figure 13 shows the essential routes involved. It takes place in all land plants and is an essential part of the plant's nutrition.

(a) *The need for water.* Water is an essential plant nutrient as it provides the hydrogen for the reduction of carbon dioxide to carbohydrates during photosynthesis. Besides this, water acts as the transporting medium for all the substances that are moved about inside the plant, and it also provides support for non-woody tissues, and it cools the plant by evaporation from the leaf surfaces. If the transpiration stream is interrupted by non-availability of water or by slow conduction or over-fast loss from the leaves then wilting results. Any extensive period of wilting leads to the death of the plant.

(b) *Disadvantages and advantages of transpiration.* The whole process of transpiration is physical and it is bound to occur from the nature of the plant. The leaf is saturated and, during daylight, porous, so water is likely to evaporate into the air.

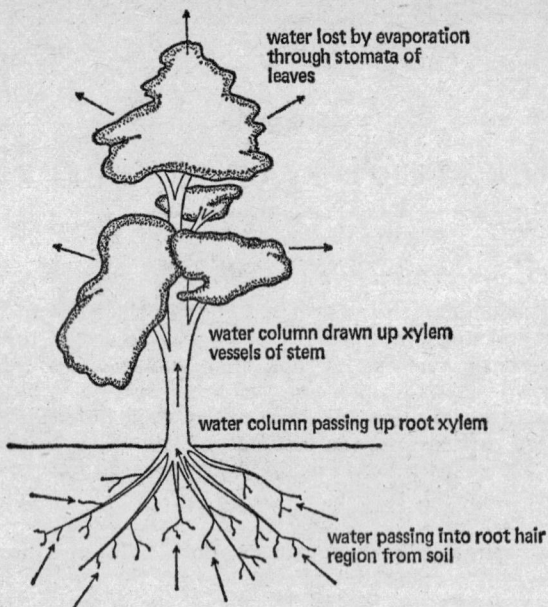

FIG. 13. The transpiration stream.

A great disadvantage of transpiration is the very large amount of water that must pass through the plant for it to remain alive. (An acre of cereal crop passes some 500,000 gallons of water through its leaves during the growth period and an acre of deciduous woodland many times more.)

One of the vital stages in the successful colonisation of the land by plants was the ability to control this water loss and yet to develop large photosynthetic surfaces.

The advantages of transpiration have been mentioned above and it is clearly essential if the plant is to remain alive at all.

3. The motive force of transpiration.

(a) *The leaf and the atmosphere.* The leaf spaces between the cells of the spongy mesophyll (*see* Fig. 12), are completely saturated with water vapour, being themselves in contact with the turgid cell contents. During the hours of daylight the stomata will be open and, unless the atmosphere is also 100 per cent saturated with water vapour, there will be a tendency for water molecules to diffuse out of the leaf.

On a normal summer day the atmosphere will be some 70 per cent saturated and this means that the tendency of water to escape from the leaves is very high. In very dry climates such as deserts with relative humidities below 30 per cent the escaping force of water may be several thousand atmospheres!

(b) *Factors affecting water loss from the leaf.* These factors are basically the stomatal aperture, the anatomy of the leaf and the external relative humidity, *i.e.*:

(i) The stomata are open during the light; as 95 per cent or more of transpiration takes place through this route clearly light is a limiting factor for the rate of loss. During the night the stomata shut and water loss is very much reduced.

(ii) Some leaves have especially thick cuticles or are very hairy, others are narrow and have few and sunken stomata. All these anatomical modifications will tend to reduce water loss and are often associated with plants growing in dry regions. Such plants (*e.g.* marram grass on sand dunes) are called *xerophytes*.

(iii) Anything which affects the relative humidity of the air will also affect water loss provided the stomata remain open. Thus wind and heat decrease water content or relative humidity of the air and increase evaporation from the leaf. Alternatively cold, mist, fog, dew and still air all increase humidity, often up to saturation. In this case water will not evaporate from the leaf and the transpiration stream will cease. The escape of water from the leaf spaces is shown in Fig. 14.

(c) *Water movement from root to leaf.* As already described in the previous chapter water is drawn into the plant root hairs from the soil by osmosis. There may be a limited force pushing this water into the xylem and up the root, called *root pressure*, but it is very low in most plants.

We have seen however that water is being lost from the leaf spaces by evaporation and this causes more water to pass into the spaces from surrounding cells. This in turn causes water to be drawn by the depleted cells from the veins. The column of water they represent is very thin (less than 0·5 mm) and stretches unbroken to the root via the xylem of the stem. Such a thin column of water has great tensile strength and as it is drawn upwards from above remains unbroken.

This force drawing water up the plants is suction and so strong are the columns of water being pulled up that pressures in excess of 300 atmospheres would be required to break them. This means water could, at least in theory, pass up a tree some 10,000 feet in height.

(d) *Wilting*. The green plant will under normal conditions have a complete column of water from the highest leaf to the deepest root. During darkness this column is virtually stationary but by day, with the opening of the stomata, it

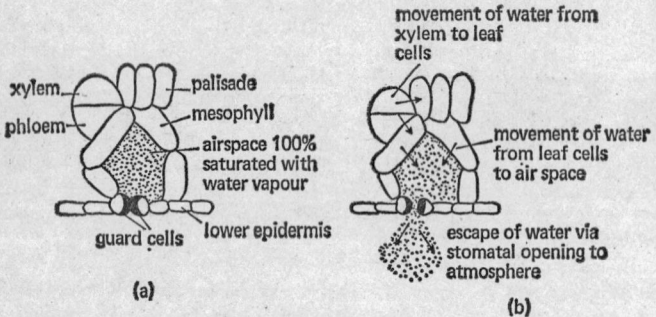

FIG. 14. The role of stomata in control of transpiration. In (a) water is contained in the leaf during darkness, and (b) shows how water is lost to the air in light whenever the relative humidity is less than 100%.

moves as the transpiration stream. All the living cells of the plant take water from this stream and are dependent upon it, being maintained in a state of turgor.

If at any time the rate of evaporation from the leaves exceeds the rate at which water can be supplied to the leaves the plant wilts. This is because water lost from the stomata cannot be replaced as fast as it is lost so the mesophyll and other cells of the leaf lose their turgor. They become plasmolysed and the leaf collapses, or wilts. Temporary wilting is common near noon on a hot summer's day but if it is maintained for a number of hours and the cells lose more than 15 per cent of their water they are likely to die.

PHOTOSYNTHESIS

4. General aspects of photosynthesis. We now have sufficient knowledge of the anatomy of the green plant and its
means of conduction to study the essential part of its nutrition
—the process of photosynthesis.

The leaf is thin and it has a large surface area so that it is
easy for light to pass through. It is also green due to the chloroplasts and these absorb a part of the sun's energy in the form of
light. Within the chloroplasts the light energy is turned to
chemical energy and this is the driving source for all subsequent
reactions.

Until a few years ago it was customary in introductory texts
to represent the process of photosynthesis by the following
reaction:

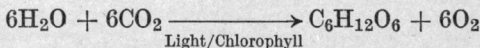

$$6H_2O + 6CO_2 \xrightarrow[\text{Light/Chlorophyll}]{} C_6H_{12}O_6 + 6O_2$$

In fact this simple equation does represent the initial and
final products of the process but it tells us very little of the
mechanism and indeed in some respects is highly misleading.
Much more is known today about this all-important process,
and the unravelling of its details has been a major advance in
biology.

From many pieces of evidence and especially from the effects
on the rate of photosynthesis of different conditions of light
and temperature and carbon dioxide concentration it has been
shown that photosynthesis is essentially a two-stage process.
These two stages are called the *light reaction* and the *dark
reaction*. (*See* Chapter VI for relationship between photosynthesis and respiration in plants.)

5. The light reaction involves the action of light on chlorophyll, the hydrolysis of water and the capture and temporary
storage of its hydrogen and release of the oxygen. This
happens as follows:

(*a*) *The role of chlorophyll.* Chlorophyll being green ab-
sorbs the red and blue parts of the spectrum. It is not the
only light-absorbing pigment in leaves, the yellows of
autumn show the colour of carotene which is normally
masked by the dominant greens.

All the pigments of the leaf operate together in the capture
of light and its subsequent transformation into chemical

energy. This energy is stored partly as high-energy phosphate or ATP (*see* Chapter VII, 9) and partly used at once to break up water in the chloroplasts into hydrogen and oxygen.

$$H_2O \rightarrow 2H + O$$

This is a very endothermic reaction (Chapter VI) and takes a lot of energy.

(b) *Release of oxygen.* The oxygen released from the water is produced in greater quantities than the amount required by the respiration of the plant and it therefore tends to diffuse out from the photosynthesising cells via the air spaces and stomata to the atmosphere. The light re-action is thus associated with release of oxygen from the leaf and capture and retention of the hydrogen.

(c) *Fate of the hydrogen.* The hydrogen released from the water during the light reaction is extremely reactive and is linked to a special hydrogen carrier molecule in the chloroplasts. In this form it is more stable and can be contained until utilised in the reduction of carbon dioxide which is the important part of the dark reaction.

6. The dark reaction involves some very complex cyclic stages but basically it results in the fixation of CO_2 to an organic molecule. The detailed pathways of the dark reaction were first worked out after the last war using the radioactive isotope of carbon (^{14}C) by which intermediate stages could be labelled. The dark reaction is as follows:

(a) *Carbon dioxide pathway.* The CO_2 which has diffused in through the stomatal openings and across the air spaces of the spongy mesophyll goes into solution at the cell surface and finally reaches the chloroplasts.

(b) *The incorporation of carbon dioxide.* CO_2 combines with a five-carbon compound substance which is also taking up the hydrogen from the hydrogen carrier (*see* above). This produces not a six-carbon compound, as might be expected, but two molecules of a three-carbon compound called *Phosphoglyceraldehyde* or *PGA*. A number of things can happen to the PGA.

(c) *The possible fates of PGA.*

(i) *Carbohydrate formation.* Perhaps the most obvious destination of the three-carbon PGA is to condense into molecules of six-carbon sugar such as glucose, $C_6H_{12}O_6$. In the sense that this is the major route, carbohydrates can be considered to be the main product of photosynthesis and this is implied in the original simple equation. Glucose in its turn condenses to maltose

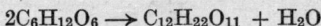

$$2C_6H_{12}O_6 \rightarrow C_{12}H_{22}O_{11} + H_2O$$

and many glucose units combine to form starch, a storage polysaccharide. This is such an extensive and typical product of photosynthesis that starch testing a leaf is the way in which the occurrence of photosynthesis is demonstrated by experiment.

Cellulose, the material of the cell wall, is another polysaccharide that can be built up from glucose units.

(*ii*) *Fat formation*. Three-carbon PGA may also suffer loss of oxygen and condense together long chains of carbon and hydrogen forming fats and oils. These can be important storage products (*e.g.* in groundnuts and olives), but fats are vital to membrane formation in all plant cells.

(*iii*) *Protein formation*. Nitrates and other ions will have been brought up to the leaf cells in the xylem and here they can be combined with the basic three-carbon PGA to make various amino acids which in turn can condense to form proteins or nucleic acids.

(*iv*) *Respiration*. All the fates of PGA described above involve the use of energy, for it can be seen that they are all building reactions (*see* Chapter II). The energy for all these reactions is provided by the breakdown of PGA via respiration pathways. These are described in more detail in Chapter VI but briefly they result in ATP formation and the production of CO_2 and H_2O wastes.

7. Summary of photosynthesis and subsequent reactions.
The information in the above sections is usefully summarised in the schematic diagram overleaf.

8. Conditions affecting the rate of photosynthesis.
The rate of photosynthesis in green plants is always limited by one factor or another. The major limiting factors are as follows:

(*a*) *The leaf*. An immature or senescent leaf will not have a high rate of photosynthesis because of a number of factors concerning the biochemistry of the cells. Chlorophyll might act as a limiting factor in such a leaf or in one that was deprived of the essential elements iron and magnesium which are required in its synthesis.

SUNLIGHT

light reaction   ↘ CHLOROPHYLL

Generation of CHEMICAL ENERGY·

Water ←- - - → Hydrogen + Oxygen (released)

HYDROGEN CARRIER

dark reaction   CARBON DIOXIDE + HYDROGEN + 5 Carbon Compound

2 Molecules 3 Carbon (P.G.A.)

GLUCOSE   FAT   AMINO ACID   via respiration
MALTOSE SUCROSE   to $CO_2 + H_2O$
           PROTEIN   and energy to
STARCH CELLULOSE   power other
               building
               reactions

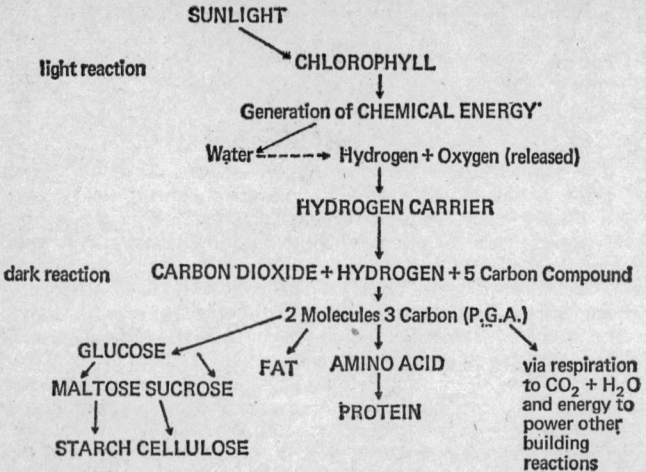

(b) *Light* is commonly a limiting factor although different plants become "light saturated" at very different intensities. Thus a shade plant such as holly may be photosynthesising at its maximum rate at a light level which would produce very little result in a sun plant such as an oak tree. On the other hand the sun plant can utilise much higher light intensities than the shade plant and where both are in optimum light intensities they will photosynthesise at a greater rate.

(c) *Carbon dioxide*. There is only some 0·04 per cent of this in the atmosphere and on a still day dense vegetation will reduce the local level almost to zero. Plants can use up to 1 per cent CO_2 and therefore normally work far below their optimum for this particular raw material.

(d) *Temperature*. The light reaction of photosynthesis is not affected by temperature but the dark reaction consists of a number of enzyme stages whose rate tends to double for every 10°C rise. This only applies during their working range (*i.e.* some 10°–40°C); outside these ranges enzyme reactions, and therefore photosynthesis, are generally inoperative.

TRANSLOCATION

9. What is translocation? In the preceding sections we have
een how the raw materials of carbon dioxide and water and of
nineral ions become converted into organic molecules. These,
nd especially starch, may find temporary storage in the photo-
ynthesising cells, but sooner or later they will be transported
way to other parts of the plant where they are utilised. This
process of transport of organic products is termed *translocation*.

10. The phloem as a transport system. The elements that
make up the phloem and their distribution have already been
described (*see* Chapter III). It will be recalled that phloem
tubes are living and make up a part of the vascular system of
leaf, stem and root.

While it is not certain how organic molecules are passed
along the interconnected phloem tubes of the plant it is clear
that these are the means of their transportation. Destruction
of the phloem by one means or another leads to accumulation
of organic substances on the producer side of the junction.

11. Sinks and sources. It is convenient to consider the
leaves which are producing organic substances as sources and
the organs to which these substances are transported as sinks.
Growing tissues would thus be sinks as would storage organs
and fruits and seeds.

In the whole life of a plant however a seed (as it germinates)
may act as a source, as may the rhizome, or tuber, or bulb, or a
perennial plant. From such storage organs food is passed out
to the meristems to allow new growth to occur in spring.

An important fact about translocation in the phloem is that
it takes place always from a source to a sink and it does not
matter if this is with or against gravity. In either case the
movement of the food is thousands of times as rapid as the rate
that might be expected from normal diffusion. Translocation is
an active process and involves the use of respiratory energy.
Dead phloem will not transport food, and so the whole process
contrasts with transpiration which is purely a passive physical
movement along dead and hollow vessels.

A summary diagram of the sources and sinks of the translocation substances and their utilisation in different parts of
the green plants is shown in Fig. 15.

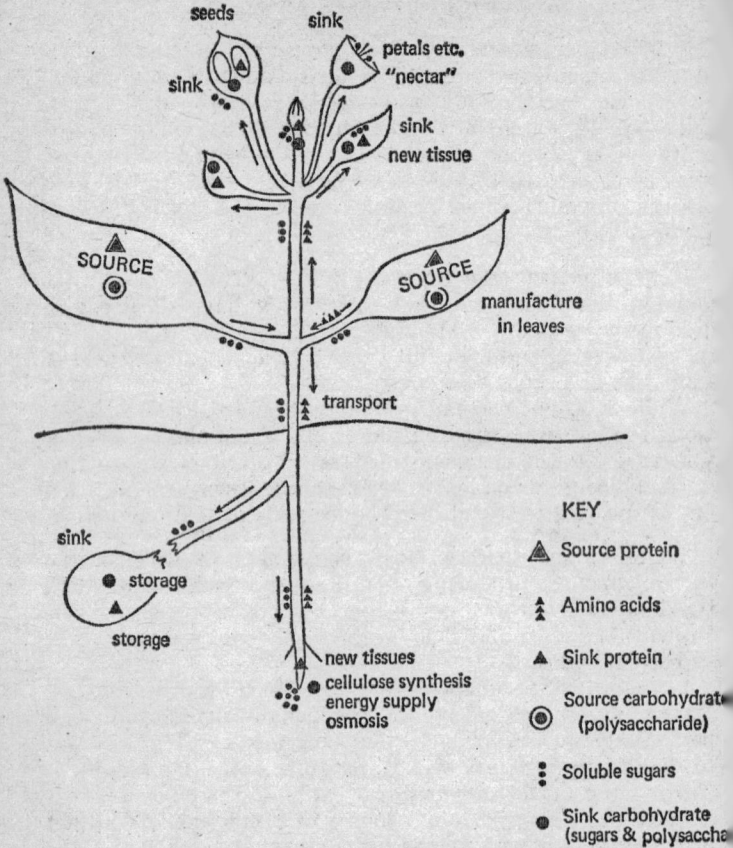

FIG. 15. Translocation in the green plant.

PROGRESS TEST 4

1. Why do plants require water? (2)
2. What is meant by transpiration? (2, 3)
3. What factors affect the rate of transpiration? (3)
4. Why do plants wilt? (3)
5. What happens in the light reaction of photosynthesis? (5)

6. What is the importance of PGA? (6)
7. What possible products may ultimately result from photosynthesis? (7)
8. What conditions affect the rate of photosynthesis? (8)
9. Where does translocation occur? (10)
10. Distinguish between a sink and a source. (11)

EXAMINATION QUESTIONS

1. The following readings were obtained on the uptake of water by a transpiring shoot:

	LIGHT	LIGHT AND FAN	DARK
Mg water taken up per minute	4	9	0·2

Comment on these results. (*after A.E.B.*, 1970)

2. Describe how carbon dioxide taken in at the leaf of a potato plant may become incorporated as part of a starch molecule in the potato tuber. (*after O. & C.*)

CHAPTER V

THE NUTRITION OF THE MAMMAL

1. Introduction. The nutrition of the green plant has been described in the previous chapters and it will be recalled that the key feature of the plant's nutrition was its ability to synthesise its requirements from very simple substances. This form of nutrition is termed *autotropic*.

Animals, on the other hand, are unable to synthesise C, H, and O compounds from carbon dioxide and water and their powers to make specific amino acids, vitamins and other essential substances are also very limited. Animals therefore depend on plants for their food and this type of nutrition is called *heterotropic*.

Following the normal syllabus requirements the nutrition of mammals and specifically man will be described in this chapter.

DIET

2. The concept of a balanced diet. The body of a man is made up largely of water (67 per cent) with an organic content of carbohydrates, fats and proteins. Besides these, much smaller amounts of vitamins and variable amounts of mineral salts complete the total composition.

A diet which contains all these items in sufficient amounts for the body to remain healthy and to grow and replace worn-out tissue is termed a "balanced" diet. If specific items are lacking from the diet then the body will develop deficiency diseases. Clearly such diseases are not infectious and can be cured by taking the missing nutritional substance. Where, however, this is not available certain of these conditions may be fatal.

3. Human diet. Some of the chemical characteristics of the substances that make up our bodies and hence our diets have been described in Chapter II. Some further comments on these as they relate specifically to man are appropriate in this section.

(a) *Carbohydrates* are the staple ingredient of our diet in the form of starch. It is commonly eaten as cereals (including flour and its products), as potatoes or as rice. Starch is the usual storage substance of plants, in overwintering organs such as tubers as well as in seeds. Peas and beans contain 60 per cent dry weight of starch, and animal starch or glycogen is a major constituent of meat, and liver.

Carbohydrates are also eaten in the form of sugar, lactose being present in milk and milk products and sucrose being refined from beet and cane.

In whatever form they are eaten carbohydrates contain about 17 joules/g and they provide the major source of energy for the metabolism of the body. Carbohydrates are stored in our muscles and in our liver in the form of glycogen, and a well-fed human will have a sufficient store to last him for many days provided he has adequate water.

(b) *Fats.* It is necessary for man to take in some quantity of fat each day because the fat soluble vitamins A and D can only be assimilated if fat is present in the gut. Besides this, the membrane structures found in all cells contain fats, and the body also produces oily secretions from the sebaceous glands of the hair follicles. In connection with his comparatively hairless state man uses a store of fat under the skin to provide insulation against the cold.

Fats are an alternative or subsidiary food for energy production and have at least twice the joule value of carbohydrates, weight for weight.

Animal fats in our diet may be derived from milk products such as butter and cheese, from meat such as bacon, or from fried foods. Vegetable fats are eaten as they make up a proportion of margarine, and also various cooking oils.

(c) *Proteins.* These are another essential of diet for the growth and repair of the body. Proteins consist of some twenty different amino acids and the body can make only half of these by "chemical manipulation" of the amino-groups. This means that not only a basic amount of protein has to be eaten, but also that protein rich in these essential amino acids has to be included in the diet.

Such protein is found in meat, fish, eggs, milk, and to a lesser extent in plant products, especially seeds. Deficiency in the correct protein leads to a disease called *kwashiorkor*, unfortunately all too common in the poorer parts of the world.

Should an excess of protein be eaten, and this is the general rule in the wealthier parts of the world, the nitrogen (and sulphur and phosphorus where these are present) can be removed by the liver and the residue stored up as glycogen for respiration.

(*d*) *Vitamins* (*see* also II, 8 which discusses enzymes and vitamins) were discovered some fifty years ago and since that time very many have been described. They are defined as substances that are required in very small quantities to maintain the health of the body. Unfortunately they cannot be stored so a constant supply is required. The main vitamins for man are as follows:

(*i*) *Vitamin A.* This has general importance in normal growth and seems to determine whether the body retains and uses proteins or breaks them down and excretes them. It is also used to make the pigment, visual purple, in the retinal cells of the eye so that a deficiency leads to deterioration of sight. Finally vitamin A is important to membrane formation and for the normal development of the epithelial cells that line the inner surfaces of the body. As these are our first line of defence against infection, one of the symptoms of deficiency of the vitamin is the susceptibility to disease. The vitamin is found in liver, milk and meat.

(*ii*) *Vitamin B* is a complex of a number of different chemical substances. The majority of these, such as thiamine and nicotinic acid, act as intermediate enzymes in tissue respiration and in the final breakdown of lactic acid, to yield energy and carbon dioxide and water. If these vitamins are lacking the body accumulates poisonous intermediate products of respiration and these cause a variety of unpleasant symptoms including paralysis.

Specific names are given to the deficiency diseases resulting from lack of vitamin B. These are *pellagra* and *beriberi*. Excellent sources of the vitamin are milk and yeast.

(*iii*) *Vitamin C* is a fairly simple chemical called *ascorbic acid*. It is not required by many mammals (*e.g.* rats) but is necessary for man. The vitamin appears to be important to the normal structure and function of the blood capillaries and where it is lacking these tend to "leak" blood into joints and other tissues. Bleeding also occurs readily from the gums and the gut lining.

The deficiency disease associated with lack of this vitamin is called *scurvy* and was very common in the eighteenth century on board ships. Very simple but effective experiments showed that small amounts of fresh fruits (especially oranges, lemons and limes) would cure and prevent the disease. In fact Vitamin C is present in many fresh fruits and vegetables such as brussels sprouts, but it is water soluble, destroyed by cooking and not very stable.

A few years ago it was considered that Vitamin C was important in building up disease resistance but this has never been demonstrated with certainty.

(*iv*) *Vitamin D* is a fat soluble chemical called *calciferol*. It may be made within the skin by the action of sunlight but is also present in meat, eggs, fish and milk among other foods. The vitamin is essential for the uptake of the two minerals calcium and phosphorus (in the form of their respective ions) from the gut and into the bloodstream. Thus a lack of the vitamin produces the disease called *rickets* where the bones become soft and bent and the teeth decay very easily.

A supply of Vitamin D is especially important to pregnant women and to young children so that the skeleton and teeth will develop in a normal manner.

(*v*) *Other vitamins*. There are many other vitamins that have been described, but a normal diet (at least in the better nourished parts of the world) is most unlikely to be deficient in them. It has become clear that vitamins play a vital role in the metabolism by acting as the key parts of certain vital enzymes. All the symptoms of vitamin deficiency disease are due to a failure of these enzymes to function. It may seem strange that the body has lost the ability to make these essential chemicals but biologists would explain this by pointing out that the vitamins are normally present in our food and therefore the necessity to make them ourselves has gradually diminished.

(*e*) *Mineral salts*. Here again a balanced diet will contain the small amounts of the mineral salts that are needed for a variety of purposes in the body. The full list is very long and most of them are never deficient enough to cause specific diseases. It seems that a craving for a certain type of food is nature's way of telling us that we need a particular salt (or indeed vitamin). Some work done on infants who were just old enough to feed themselves indicated that they tended to pick out a more or less balanced diet from a wide range of foods presented to them. Some of the major salts and their uses are shown below:

Mineral	Form absorbed	Use in the body	Symptom of deficiency
Iron	Fe	Haemoglobin Respiratory enzymes	Anaemia
Calcium	Ca	Bone and tooth manufacture Blood clotting Muscle contraction	
Phosphorus	PO₄	Bone and tooth manufacture Energy release Nerve conduction	
Sodium	Na	Important blood and tissue salts	
Potassium	K	Nerve conduction, muscle function	
Chlorine	Cl	Stomach acid, nerve function	
Fluorine	F	Traces in teeth	Susceptibility to decay
Iodine	I	Hormone thyroxine	A form of goitre

(f) *Water*. This sounds so ordinary that we tend to forget that it is perhaps the most essential of all the items of our diet. At the least, the body will use some two litres of water every twenty-four hours. These uses are made up as follows:

(i) *For maintenance of our state of hydration*. All the re-actions of the body take place in solution: the uptake of oxygen in the lungs, of food from the gut, chemical reactions within cells, the transport of food and waste and hormones in the blood. These and indeed all reactions and activities of the body take place in solution. Where dehydration takes place metabolism ceases to function and death must result.

(ii) *For excretion* (see also Chapter VII). Elimination of the toxic products of nitrogen metabolism, *i.e.* ammonia and urea, must take place in dilute solution. Excess of mineral salts taken in our food must also be removed in this way. It is not possible to produce either very strong or semi-solid urine and a constant supply of water passing via the gut to the bloodstream to the tissues and thence back to the blood and finally to the kidneys is essential for waste elimination.

(*iii*) *For temperature regulation* (*see* also Chapter VII). The secretion of sweat is the main cooling mechanism of the body and its means of maintaining a constant temperature. If heat generated in the tissues is not constantly passed by the blood to the skin and there lost by the evaporation of sweat, explosive temperature rise occurs and again death will result.

Obviously in hot weather or warm climates a great deal more water than normal will be required for this function.

(*g*) *Final considerations on the balanced diet.* Besides all the classes of foods described above it is clear that the diet must contain sufficient energy to supply the needs of the body, *i.e.* we must have enough to eat. The quantity of food required is related to the joules of energy used each day (1 calorie = 4·2 joules). First there is the basic minimum to keep our reactions (heartbeat, breathing movements, etc.) going and to maintain our temperature. After this we must add the number of joules required for additional activities, such as brain operation and movement. These, and especially the latter, depend very much on our age and the sort of jobs we may have. An adult office worker who takes little exercise will probably require only some 14,000 J/day while a miner or labourer could use over 20,000 J in the same period.

If a man was to take a perfectly balanced diet in the form of food concentrates and vitamin pills and water he would still not be getting a really satisfactory intake. The gut needs to work on the tough indigestible fibrous material in plants to keep its tone (just as the muscles in the limbs need to be used to retain their power). These indigestible elements in food are called *roughage*. They obviously make up a good deal of the final faeces eliminated by the gut but during their passage through our alimentary canals they ensure that the muscles retain their normal capacity to bring about peristalsis upon which healthy digestion depends.

THE TEETH AND THE MECHANICAL BREAKING UP OF THE FOOD INTO SMALL PARTICLES

4. Composition of the teeth. All types of teeth have the same basic anatomy (*see* Fig. 16). Sections of a human tooth show the following layers and structures:

(*a*) *The enamel* is the outer layer of the tooth and includes the part that can be seen running a short distance under the gum. Enamel is a particularly hard crystalline form of

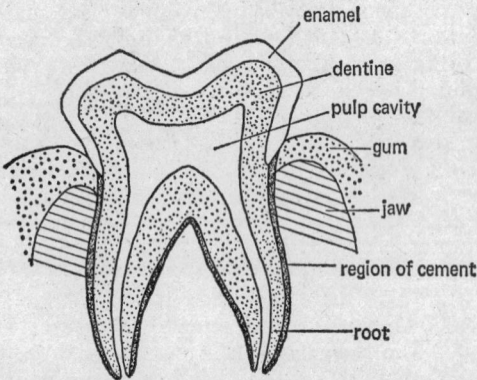

FIG. 16. The human tooth.

$CaCO_3$, harder even than bone (thus dogs can crunch up bones without breaking their teeth). It is a secretion and is not a living substance nor is it permeated by canals. Axes of the crystals are orientated to give the most powerful cutting action while preventing splitting. There is now very convincing evidence that minute quantities of the element fluorine increase the hardness and durability of our dental enamel.

(*b*) *The dentine* is the second layer of the tooth and is more extensive than the enamel. It is a hard substance related in composition to bone and tiny canals ramify through it from the living content of the pulp cavity below. The substance of the dentine is very slowly exchanged, new mineral salts being incorporated as older ones are lost, and in man, equilibrium of growth and loss is so balanced that the mature tooth does not change size.

Unfortunately the dentine is susceptible to decay once the acids of decaying food and bacteria have dissolved through the protective enamel. One day it may be possible to immunise people against those specific bacteria which cause dental caries in the same way in which they can be immunised against other bacteria such as those that cause TB or diphtheria.

(c) *The pulp cavity* lies within the centre of the tooth and is full of blood vessels and the cells which make the teeth as well as nerve endings. Connection is made to the underlying vessels through the hollow roots which are embedded in the jaw. Man has very narrow roots to his teeth compared with herbivorous animals such as horses whose teeth are continually growing.

(d) *The cement* is another bone-like substance which lies between the roots and the bone of the jaw. As its name indicates it hardens to lock the whole tooth firmly in position.

5. Types and arrangement of human teeth. Possibly one of the important features accounting for the success of the mammals was their ability to use all sorts of natural substances for food. This ability is associated with having different types of teeth to do different jobs. (*See* Fig. 17 for arrangement of human teeth.)

In the herbivores such as cows, horses and sheep etc., highly specialised molars are found which can grind hard

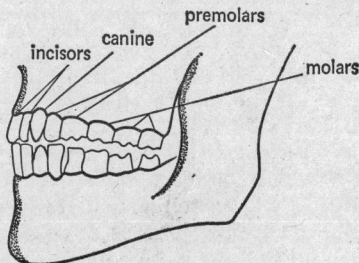

FIG. 17. Human dentition.

cellulose plant foods into minute particles to assist their digestion. In carnivores such as the cat family there are long stabbing teeth for killing their prey and the molars are sharp for tearing meat. Man, however, like pigs and bears, is an omnivore and his teeth are not highly specialised for any particular diet. The human dentition is as follows:

(a) *The incisors* are the front teeth. There are four in both top and bottom jaws. The incisors are chisel-shaped along the front of the jaw and allow us to bite our food.

(b) *The canines* or eye teeth are much reduced in man. They lie behind the incisors and there are two in each jaw. They have no specific function but assist in the general chewing action of the whole set. Both canines and incisors are single rooted.

(c) *The premolars* are behind the canines and there are four in each jaw. The cusps at the crown of these teeth are rounded and arranged so that those from the bottom jaw exactly fit those at the top. Premolars act with the molars in providing an efficient chewing action which reduced the lumps of food bitten off by the incisors into particles small enough to swallow. The chewing action also ensures thorough mixing of the food with the salivary juices.

(d) *Molars* are the last teeth at the back of the jaw and there are six of these on both top and bottom. They are bigger than the premolars but function in exactly the same way, providing the main chewing action of the set. Both premolars and molars have more than one root.

(e) *The dental formula* is a quick means of expressing the numbers and types of teeth in each jaw. For each type of tooth the letters are given—*i.e.* I = incisors, C = canines, Pm = Premolars and M = Molars—and then the number in upper and lower jaw. As the teeth are symmetrical, only half the jaw is given, but the final figure is made equal to the total number in the mouth. Thus for man the dental formula is $I\frac{2}{2}C\frac{1}{1}Pm\frac{2}{2}M\frac{3}{3} = 32$

(f) *The jaw muscles and chewing action.* The ramus of the lower jaw in which the teeth are set is extended into an upwards running condyle and articulates with the upper part of the skull. The jaw joint is a loose one (compared with the elbow) and allows both upwards and downwards as well as lateral movements. There are two major jaw muscles, both concerned with pulling the bottom jaw up towards the top, and together they provide the power for the biting and chewing actions which are the functions of our teeth.

Like the teeth themselves our jaw muscles are much reduced from those of our ape-like ancestors. Because of his unnaturally soft and sugary diet modern man seems much more subject to tooth decay than do other mammals.

THE ALIMENTARY CANAL AND THE DIGESTION OF FOOD

6. The alimentary canal (*see* Fig. 18) is a tube some twenty-two feet in length, running from the mouth to the anus. It is divided into a number of different functional areas: the digestive region, the assimilatory region, and the waste

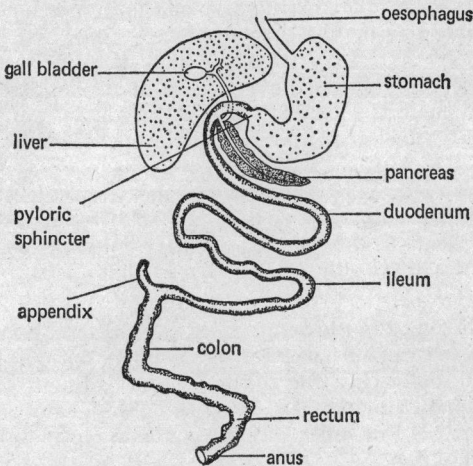

FIG. 18. The human alimentary canal.

consolidation and elimination regions. Apart from this long glandular tube there are two major organs associated with the processes of digestion and assimilation of food, the pancreas and the liver.

In the following sections each region of the gut will be

described together with its secretions and functions while a summary of the whole process of digestion and assimilation will bring together the essential information.

7. The buccal cavity, or mouth cavity, contains the teeth whose structure and function have been outlined above. In addition to the mechanical breakdown of food materials, secretions of three pairs of glands empty into the buccal cavity and become closely mixed with the food during chewing. The collective name for these juices is *saliva* and their individual composition and activities are given below.

(*a*) *Mucus and water.* Dry foods are moistened by these substances to improve efficiency of mixing and to allow eventual ease of swallowing. A man who is frightened and whose saliva has dried up (*see* Chapter IX) would find it almost impossible to swallow a dry biscuit.

(*b*) *Amylase.* This is the first of the many digestive enzymes and acts on the carbohydrate starch, hydrolysing it to the double sugar maltose, the first stage in its digestion. Of course the food does not remain in the mouth for very long and there is a further secretion of amylase lower down in the gut which gives this reaction a second chance to take place.

(*c*) *Alkaline.* Like all enzymes, the various digestive ones operate most efficiently at different acidities or alkalinities, and amylase works best in slightly alkaline conditions: hence the saliva contains some bicarbonate.

8. Swallowing and the oesophagus. As the food is swallowed from the back of the buccal cavity a reflex action closes the entrance to the trachea and to the palate (leading to the nose). Thus the food can only pass down the oesophagus or tube to the stomach. If this reflex goes wrong, as sometimes happens, it causes choking.

The oesophagus, like the rest of the gut, has two sets of muscles: one running up and down called the *longitudinal,* and the other going round called the *circular.* The action of these two sets of muscles causes peristalsis, the means by which a given lump of food is pushed along from the mouth towards the stomach. This is a very positive action and is not just due to gravity; it is quite possible, though perhaps not

,dvisable, to eat and drink standing on one's head. The food
hen enters the stomach.

9. The stomach. This is a muscular bag-shaped organ with
considerable powers of distention and contraction. Besides
he normal sets of muscles, there is a further layer which
gives the stomach the necessary force to churn the food and
complete the mechanical action.

A glandular mucosa lines the walls of the stomach, into
which are sunk many hundreds of thousands of gastric pits.
It is from the cells in these pits that the active gastric juice
is secreted by the stimulus of hunger or the presence of food
in the stomach. The gastric juice consists of:

(a) *Water and mucus,* which have similar functions to the
saliva except that the mucus probably provides additional
protection to the stomach lining from its powerful acids and
protein-digesting enzymes.

(b) *Pepsin* is the main enzyme of the gastric juice. It is
a very active protein-splitting enzyme hydrolysing the
huge molecules of the protein into much smaller polypeptides.

(c) *Hydrochloric acid* at 1/10 Normal (Normal is gram
molecular weight of a chemical/litre) is secreted and makes
the stomach contents very acid. It may bring about the
digestion of some protein and digestion of other foods on
its own and may also kill bacteria that enter the stomach
with the food, but there can be no doubt that its major
function is to work with pepsin. It activates the enzyme
and provides the acidic medium in which it is most efficient.

(d) *Rennin* is present in the gastric juice of young
mammals and starts the digestion of milk by precipitating
the insoluble casein from the soluble milk protein caseinogen.

After the food has been churned and acted on by these
juices for several hours a major emptying of the stomach
takes place due to the relaxation of the pyloric valve which
separates it from the duodenum. In the duodenum the process
of actual digestion is completed and this is partly accomplished
by the secretions of the two major organs that empty into it,
the liver and the pancreas. The secretion of these organs will
be described first.

10. The liver is the largest organ in the body and has many functions. Later in this chapter we shall see how it stores and changes many of the foods assimilated by the gut, but in the present context the liver's secretions into the gut will be described.

Blood flowing through the many thousands of channels of the liver is in close contact with minute ducts of the bile system. The bile flows along these ducts to the gall bladder and is periodically emptied from it via the bile duct to the duodenum.

(a) *Composition of bile.* Bile is a green fluid that owes its colour to a breakdown product of the red corpuscle haemoglobin. It also contains alkaline ions and bile salts.

(b) *Function of bile.* The pigment substance of bile is probably best regarded as an excretory product. It certainly colours the faecal matter and may also stimulate intestinal movements. The alkaline nature of bile helps restore neutrality to the duodenal contents to allow the most effective conditions for enzyme function. The bile salts have a detergent action and emulsify (that is, make into very tiny droplets) any fats present in the duodenum. This assists their subsequent hydrolysis by the enzyme lipase.

11. The pancreas is situated in the loop formed by the duodenum, and empties its digestive secretions via the pancreatic duct which in turn joins the bile duct, thus leading to the duodenum.

Pancreatic juice contains a number of enzymes that are important in digestion and like the bile it is alkaline. This again helps to reduce the acidity of the duodenal contents. Enzymes of the pancreas are:

(a) *Lipase*, which is a fat-hydrolysing enzyme which converts fats into smaller units called fatty acids and breaks off the glycerol molecules which bind fatty acids together. This is the only enzyme involved in fat hydrolysis.

(b) *Amylase*, which acts in the same way as the amylase of saliva, breaking down starch molecules to the double sugar maltose. This, it will be remembered, gives the gut a second chance of starch hydrolysis.

(c) *Trypsinogen* is an inactive form of the protein-digesting enzyme trypsin which is activated by a duodenal

secretion called *enterokinase*. These protein-digesting enzymes are very powerful and the intestinal walls are themselves of a protein nature. Perhaps it is not fanciful to suppose that our intestine keeps the active form of the enzyme
to a local area?

12. The duodenum is a tube about twelve inches long into
which bile and pancreatic juices are poured. It also produces
a number of other enzymes from its glandular walls which
lead to the final hydrolysis of all types of food substance.
The enzymes are as follows (*see* II, 3):

(*a*) *Maltase*, which converts malt sugar (maltose) into
two molecules of glucose, which is a single sugar.

(*b*) *Sucrase*, which converts any cane sugar (sucrose)
which is present into a molecule of glucose and a molecule
of fructose.

(*c*) *Lactase*, which converts any milk sugar (lactose) present into two molecules of galactose which is a single sugar
closely related to glucose.

(*d*) *A proteolytic complex* formerly called *erepsin* but now
known to consist of a number of different enzymes. These
hydrolyse polypeptides to peptides and amino acids and
also break up nucleic acids into small products that can be
assimilated.

At the end of its time in the duodenum the food will have
been completely hydrolysed, so that all carbohydrates are
present as single sugars, all fats either as fatty acids and
glycerol or as very small droplets, and all proteins as amino
acids. This represents the end of digestion.

ASSIMILATION OF FOOD INTO
THE BLOODSTREAM

13. The ileum. The part of the gut which follows the
duodenum is called the *ileum*. It is several feet in length and
its inner surface is covered with little fingers or villi which
greatly increase its surface area. The ileum has a rich arterial
blood supply (from the mesenteric artery) and is drained by
the hepatic portal vein which in turn leads to the liver. Within
the villi and ramifying through the underlying tissues are the

SUMMARY OF DIGESTION

Part of the gut	Secretion	Effect on food	Other action
Buccal cavity	Saliva	Amylase Starch → Maltose	Mechanical action of teeth

Stomach	Gastric juice	Pepsin + HCl
		    Protein → Polypeptide
		Rennin
		    Milk protein is
		    precipitated
			Churning action

Gall bladder	Bile	Emulsifies fats	—

Pancreas	Pancreatic juice	Amylase
		    Starch → Maltose
		Lipase
		    Fats: →
		    Fatty acids
		    + Glycerol
		Trypsinogen
			—

Duodenum	Duodenal juice	Enterokinase
		
		Trypsin
		    Polypeptides →
		    peptides and
		    amino acids
		"Erepsin"
		    Polypeptides →
		    peptides and
		    amino acids
		Maltase
		    Maltose → Glucose
		Sucrase
		    Sucrose → Glucose
		    Fructose
		Lactase
		    Lactose → Galactose
		    + Glucose
			—

Part of the gut	Secretion	Effect on food	Other action
Ileum	—	—	Assimilation of glucose, fructose, galactose and amino acids into blood and of fatty acids and fats into lacteals
Colon or large intestine	—	—	Assimilation of water. Consolidation of faeces
Rectum	—	—	Storage of faeces
Anus	—	—	Elimination of faeces

lacteals which are a part of the lymph system. Assimilation of foods takes place in the ileum.

(a) *Assimilation of carbohydrates.* Any molecules of single sugars, particularly glucose which is likely to be most abundant, are actively taken up at the surface of the villi and passed to the capillaries of the hepatic portal vein. From here they are transported to the liver where they are utilised in various ways (*see* **15–17** below).

(b) *Assimilation of proteins.* The hydrolysed products of proteins are assimilated in the form of amino acids as are sugars, and are also carried to the liver where they may undergo a variety of fates.

(c) *Assimilation of fats.* In the form of tiny droplets, or as fatty acids, these are assimilated via the villi to the lacteal and thence enter the lymph system of the body. They travel in the lymph as actual drops of fat and may be returned to the blood at the thoracic duct where the lymph system meets the vena cava. A smaller proportion of fat enters the blood of the villi directly.

Other foods such as vitamins and salts will be assimilated

at this stage if not before so that what leaves the ileum is indigestible matter and a good deal of water. This now enters the large intestine.

14. The large intestine, or colon. An enormous population of bacteria live in the final region of the human intestine. Some of these may actually synthesise vitamin B which can be absorbed into the blood but for the most part they have neither beneficial or adverse effects. In the large intestine the faecal matter is consolidated by the withdrawal of water and finally stored in the rectum until it is evacuated.

FATE OF FOODSTUFFS AFTER ASSIMILATION

15. The fate of glucose and other single sugars. These sugars are carried to the liver by the hepatic portal vein and there they are taken up by liver cells which line the canals through which the blood passes. The sugar is converted to glycogen, a polysaccharide carbohydrate similar to starch, and this is stored in the liver until required.

When this sugar store is required to be mobilised by the body, the interaction of the two hormones adrenalin and insulin causes its partial release into the bloodstream. The sugar, now in the form of glucose, will pass to the muscles and other tissues where it is required.

The liver acts as a major storage organ for reserve carbohydrates. What happens to the sugar in the tissues is described in the following chapter on respiration.

16. The fate of fats. These are mainly transported by the lymph from the ileum. As they pass around the body some may be laid down as storage or insulation fats under the skin or around organs such as the kidney or heart.

The lymph finally empties into the blood system as the vena cava enters the heart so that fats will eventually pass through the liver (via the hepatic artery). Here they may be oxidised to yield energy, or the molecules may be rearranged to give glycogen which will undergo the same fate as described above. Fats are important in the manufacture of cell membranes and other organelles within the cells of the body.

17. The fate of proteins. These are transported in the form of amino acids to the liver by the hepatic portal vein. Amino acids may broadly be divided into two categories, those the body cannot manufacture itself and those that it can (*i.e.* from other amino acids). The former are necessary or essential and they pass via the circulation to all the cells of the body. The latter are taken up by the liver cells and broken down so that the nitrogen and other elements are removed and the C, H, O, part is incorporated into glycogen.

This breakdown is called *de-amination* as it involves the removal of the ammonia which is a very toxic substance and is rapidly incorporated with carbon dioxide to make the much less harmful urea:

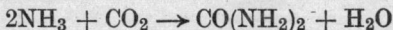

$$2NH_3 + CO_2 \rightarrow CO(NH_2)_2 + H_2O$$

This is eliminated by the kidney.

PROGRESS TEST 5

1. What is a balanced diet? (2)
2. In what foods do you find (*a*) carbohydrates, (*b*) fats, (*c*) proteins? (3)
3. What is a deficiency disease? (3)
4. What is the role of vitamins A, B, C, and D in the human body? (3)
5. Give five minerals required by the body together with their respective uses. (3)
6. Why do we need water in our diet? (3)
7. What three materials are found in teeth? (4)
8. What is the dental formula for man? (5)
9. What happens to food in the mouth? (7)
10. What secretions are found in the stomach and what are their functions? (9)
11. What are the functions of the liver? (10)
12. Name three enzymes secreted by the pancreas. (11)
13. What are the final products of digestion? (12)
14. What is the function of the villi of the ileum? (13)
15. What happens to sugar assimilated in the blood from the gut? (15)
16. What may happen to excess of protein taken into the body? (17)

EXAMINATION QUESTIONS

1. What important food substances may a man obtain from:
   (a) fish and chips
   (b) a green salad
   (c) a glass of milk
   (d) a slice of wholemeal bread and butter?
What do you understand by the term deficiency disease? Give an example of such a disease. (*O. & C.*, 1970)

2. Give an account of the digestive processes carried out in a named mammal on food from the time it enters the stomach until it leaves the small intestine. (*O. & C.*, 1968)

3. (a) What are carbohydrates? Give two examples. Why are they required in the diet of a mammal?

(b) Before carbohydrates can enter the bloodstream they must be processed. Describe the treatment they receive in the alimentary canal of a named mammal.

(c) How is the balance of carbohydrate in the body controlled? (*Cambridge*, 1969)

4. The table below shows the daily diet of women doing light engineering work in Birmingham in 1943.

Meal		kilojoules	protein (g)	fats (g)
Breakfast	7 am	2980	21	26
Snack	10 am	880	9	8
Lunch	1 pm	3490	30	27
Tea	4 pm	1010	5	9
Supper	7 pm	2390	19	21

Much of the protein eaten in the human adult diet is used as an energy source as are fats and carbohydrates. Given that 1 g of protein when oxidised in the body gives approximately 17 kJ, 1 g of fat = 38 kJ, and 1 g carbohydrate = 17 kJ.

(i) Assuming for simplicity that the protein eaten is used entirely as a source of energy, show how many kilojoules were derived from the protein content of each meal.

(ii) Do the same for the fats.

(iii) Calculate the number of joules which must, therefore, have come from the carbohydrate eaten at each meal.

(iv) Calculate the number of grams of carbohydrate eaten at each meal.

(*A.E.B.*, 1970)

5. Describe how water in the alimentary canal of a mammal is absorbed, and indicate how some of the water may eventually reach the skin and be exuded as sweat. (*Cambridge*, 1967)

CHAPTER VI

THE RESPIRATION OF GREEN PLANTS AND MAMMALS

CHEMISTRY OF RESPIRATION

1. General features of respiration.

(a) *Definition.* Respiration is the breaking down of suitable organic molecules within the cells of living organisms so that the energy they contain is made available for the other metabolic processes of life.

Whatever the "fuel" molecule utilised, the respiratory breakdown will be exothermic in nature, that is to say, it will cause the release of energy. Other life processes such as growth, secretion, movement and conduction all depend on endothermic reactions which take up energy.

Respiratory energy drives both the cell and the whole organism. Where respiratory reactions are prevented (as for example aerobic respiration is by cyanide) the organism will instantly die.

(b) *The importance of ATP.* The normal fuel molecule or respiratory substrate of cells is sugar, although it is possible for many other organic molecules to be utilised, for example fats and oils. Despite the fact that any respiratory substrate must contain a certain amount of energy it is not necessarily in the concentrated and readily available form that the metabolic processes demand.

Therefore it is clear that an important part of respiration is an intermediate stage between energy release and energy utilisation. Thus by a series of down-grading reactions the chemical energy of the substrate (say glucose) is transferred to an intermediate energy-storing molecule and this latter is used as may be required.

The most important of the energy-storing molecules is the substance ATP or adenosine triphosphate. ATP contains no less than 42,000 joules/molecule and is very readily broken down to the lower energy ADP (adenosine diphosphate), yielding its stored energy for metabolism. (If we take an analogy, ATP is to sugar what petrol is to crude oil in an engine.) The vital role of ATP in respiration may be shown as follows:

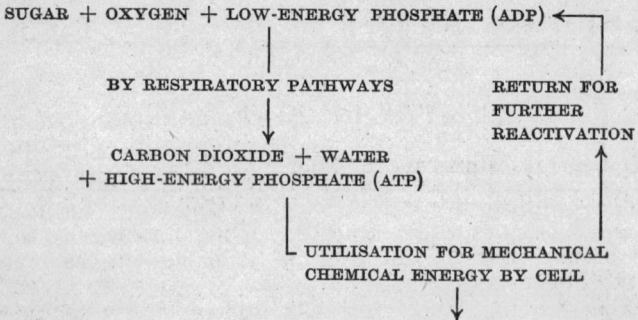

SUGAR + OXYGEN + LOW-ENERGY PHOSPHATE (ADP) ◄──────

BY RESPIRATORY PATHWAYS   RETURN FOR FURTHER REACTIVATION

CARBON DIOXIDE + WATER + HIGH-ENERGY PHOSPHATE (ATP)

└── UTILISATION FOR MECHANICAL CHEMICAL ENERGY BY CELL

(c) *The efficiency of respiration.* While the reactions of respiration are often shown as a single equation (*see* below) in fact they take place in very many stages. The overall efficiency of respiration depends on whether it takes place aerobically or anaerobically.

(*i*) *Aerobic respiration* is the down-grading of substrates by pathways that utilise oxygen. In its simplest form the reaction for aerobic respiration may be shown:

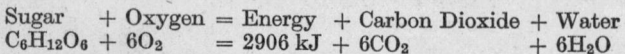

Sugar + Oxygen = Energy + Carbon Dioxide + Water
$C_6H_{12}O_6$ + $6O_2$ = 2906 kJ + $6CO_2$ + $6H_2O$

In this case it is assumed that the sugar and oxygen react directly together and that the energy is all given off in the form of heat. All these are in fact oversimplifications but the equation above remains a useful summary of initial and end products of aerobic respiration.

If 1 g of sugar is burned completely in a calorimeter 17·2 joules are produced. Within the living system only 38 ATP are made available so that this sort of respiration is approximately 54 per cent efficient.

The enzymes involved in aerobic respiration are to be found in the mitochondria of cells. Thus very active cells such as those that are rapidly dividing or secreting have many mitochondria (*see* also Chapter I).

(*ii*) *Anaerobic respiration* is the down-grading of substrates by pathways that do not utilise oxygen. Some of these pathways are identical with the aerobic process but there comes a stage where an intermediate compound with a high energy content can be broken down no further and must be excreted. Plants that are respiring anaerobically turn out alcohol whereas animals may produce lactic acid and other products. A simplified reaction for anaerobic respiration is as follows:

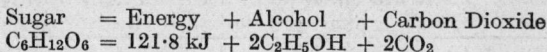

$$\text{Sugar} = \text{Energy} + \text{Alcohol} + \text{Carbon Dioxide}$$
$$C_6H_{12}O_6 = 121 \cdot 8 \text{ kJ} + 2C_2H_5OH + 2CO_2$$

Again a number of intermediate reactions are involved. It can be seen that the energy yield from this reaction is very much less and the efficiency is only 2 ATP or approximately 3 per cent. (Alcohol in concentration is inflammable and if an organism is excreting inflammable waste products it does not say much for the efficiency of its combustion!)

It might be questioned why living organisms should ever respire anaerobically. The answer is that many environments exist, as for example rotting meat and vegetation, waterlogged soil, fens and the gut of large animals where there is no oxygen available. It is essential for the saprophytes (and/or parasites that live in such places) to be able to respire anaerobically if they are to survive.

2. Temperature as a factor in respiration. All the stages involved in respiration depend on enzymes and it is a characteristic of enzyme reactions to double in rate for every 10°C rise. For those organisms, and they are by far the most common, who do not maintain a constant temperature, the respiration rate is determined by the temperature of their surroundings.

In actual fact enzyme reactions are so slow below 5°C that respiration virtually ceases. Optimum temperatures are usually around 30°C, and if the temperature is kept higher than this the enzymes eventually become denatured and the organism dies.

For birds and mammals an even temperature is maintained and this is the one at which their enzymes act most efficiently.

HUMAN RESPIRATION

3. The respiration of man. The respiration of man will be considered under three headings. First is the gaseous exchange which takes place in the lung and this is followed by transport of the respiratory oxygen and waste carbon dioxide in the blood. The final and most important part of the process is the release of energy and this takes place in the tissues themselves.

(*a*) *The lung.* Air passing in through the mouth or nostrils travels down the trachea via the bronchi to the lungs. These are made up of many millions of tiny air sacs called *alveoli* which are supplied by numerous blood capillaries. Between the air sacs there is connective tissue of an elastic nature which allows the lung volume to change. (*See* Fig. 19).

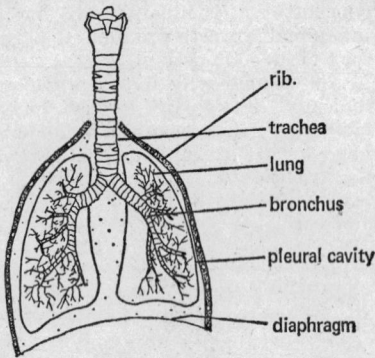

FIG. 19. The lung.

The walls of the alveoli are very thin and are moist so that oxygen and carbon dioxide readily diffuse across. When the lung is expanded its total capacity may be some five litres and a surface area of several hundred square feet is exposed to the air that has been breathed in. Cartilage rings in the trachea and bronchi prevent their collapse on breathing out.

(*b*) *Ventilation* of the lungs takes place rhythmically and is brought about by changing the size of the pleural cavity partly by means of the rib cage and also by the diaphragm muscle that separates the thorax from the abdomen. The process is described below, and the mechanism is shown diagrammatically in Fig. 20.

(*i*) *Inhalation,* or breathing in, is due to contractions of the external intercostal muscles that attach one rib to another. These have the effect of lifting up the whole rib cage, and the

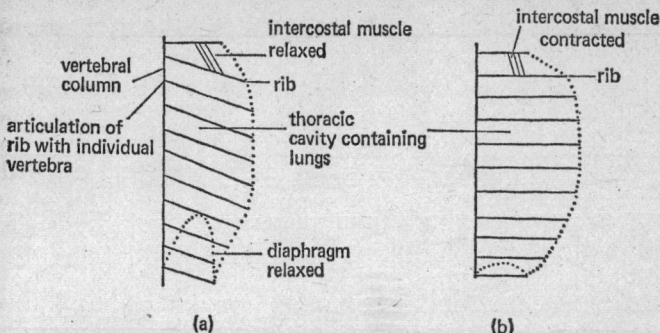

FIG. 20. Ventilation of the lung. (a) shows how, during exhalation, the intercostal muscles between the ribs are relaxed, and the diaphragm dome-shaped. Thoracic volume is minimal and air is forced out of the lungs. (b) inhalation, shows the intercostal muscles between the ribs contracted so the ribs are pulled outwards and upwards. The diaphragm contracts to a flat sheet. The thoracic cavity increases in volume, and air is drawn into the lungs.

hinges that the ribs make with the thoracic vertebrae allow this to happen. When the rib cage is raised the volume of the thorax is automatically increased hence the lungs expand as air is drawn into them.

At the same time as this is happening the large diaphragm muscle contracts and straightens out flat from its normal dome shape. This again increases the size of the pleural cavity causing air to be taken in.

Normal ventilation is of the order of 500 cc/breath and takes place some twelve times a minute. Under exertion the ventilation volume and rate will both increase.

(ii) *Exhalation,* or breathing out, is by reduction of the volume of the pleural cavity which presses on the lungs and causes air to be pushed out. The internal intercostal muscles on the inside of the ribs pull the cage down and the force of gravity helps here also. Meanwhile the diaphragm returns to its normal relaxed dome shape further reducing the thoracic volume.

4. Gaseous exchanges. Despite the tidal ventilation of the lung a continuous gradient of oxygen and carbon dioxide exists from the tissues to the atmosphere.

(a) *Oxygen* is diffused across the thin moist alveolar walls

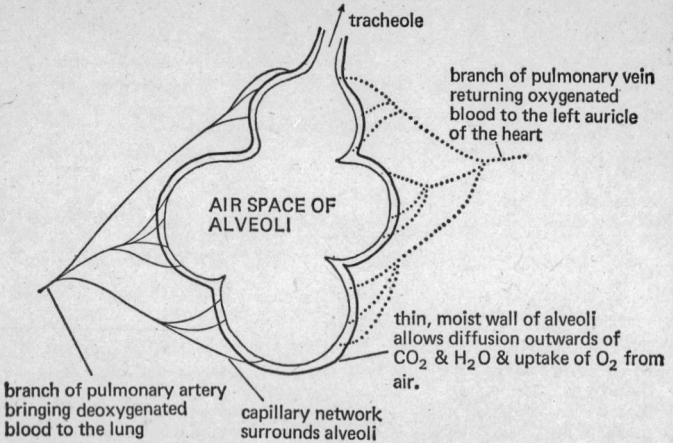

FIG. 21. A group of alveoli showing adaptations for efficient exchange of gases.

from the air within the cavity of the sac. These exchanges are shown in Fig. 21. It passes into the numerous blood capillaries that surround each alveolus and combines there with the haemoglobin of the red corpuscles. The oxy-haemoglobin so formed travels within the corpuscles back to the heart and thence to the rest of the body (*see* below).

(b) *Carbon dioxide* diffuses out of the blood which has come to the alveoli via the pulmonary artery. Combined with water as carbonic acid (H_2CO_3) it passes across the alveolar walls in the reverse direction to the oxygen.

Once in the cavity of the alveoli, CO_2 and H_2O form from the carbonic acid and these are expelled along their diffusion gradients to the atmosphere.

Compared with the gill of a fish, the human lung is not very efficient in the percentage of oxygen it removes from the incoming air. The figures are as follows:

	Inhaled air	*Exhaled air*
Nitrogen	80%	80%
Oxygen	20%	16%
Carbon Dioxide	Trace	4%
Water	Trace	Saturated
Temperature	Atmospheric	Body—37°C

5. The role of the blood in transport of respiratory gases. In order to understand how the blood functions in the transport of oxygen and carbon dioxide as well as food, wastes, heat and other substances it is necessary to interrupt this account of respiration and to consider the circulatory system.

6. The heart as a pump. The heart consists of four chambers and the left side heart and the right side heart act together but no blood can pass from one side to the other (*see* Fig. 22).

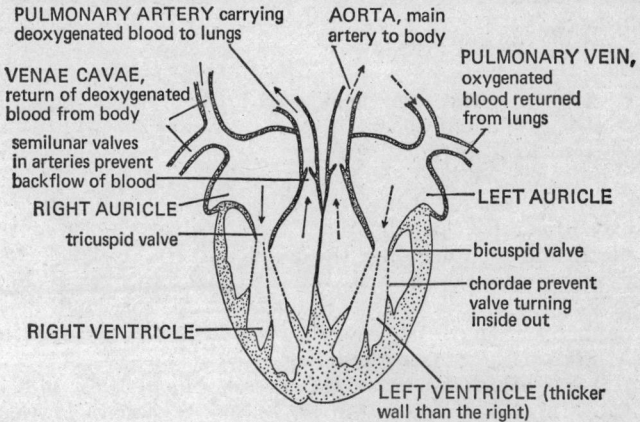

deoxygenated blood
oxygenated blood

FIG. 22. The heart and its major vessels.

De-oxygenated blood returns from the veins of the body via the great venae cavae which open into the right auricle. This in turn discharges into the right ventricle, auricle and ventricle being separated by the tricuspid valve which prevents the back-flow of blood. The right ventricle discharges via the semi-lunar valves into the pulmonary artery which leads to the capillaries of the lung.

Because the resistance of the lung is much less than that of the rest of the body the right-hand side of the heart is less muscular than the left-hand side.

After oxygenation in the capillary network which supplies the alveoli the blood returns via the pulmonary vein to the left auricle of the heart. From here it passes to the very powerful left ventricle across the bi-cuspid valve. Contraction of the left ventricle drives the blood into the aorta or main artery of the body.

The normal heart beat rate is seventy-two times a minute and approximately one gallon of blood passes through it each minute. Under exertion the beat may increase to 140 times a minute and the output may go up to eight gallons a minute.

7. Arteries, veins and capillaries. The various types of vessels that make up the circulation each have special features important to their functioning.

(a) *Arteries* are thick muscular vessels, and with the exception of the pulmonary they all carry oxygenated blood and all (including the pulmonary) take blood from the heart to the tissues. The arterial blood is at high pressure and the pulse, resulting from the rhythmical pressures of the heart beat, can be detected at the arteries. Arteries terminate in capillaries.

(b) *Capillaries* are very fine vessels of only 0·01 mm in diameter. They have thin walls and exchanges between the blood and tissues take place across these. A human body will have many thousands of miles of capillaries although not all are full of blood at any one time. All the tissues of the body, even including the bones and teeth, have capillary networks that ramify through them.

(c) *Veins.* Capillaries join up to form the veins. These return de-oxygenated blood to the right auricle of the body. The single exception to this is the pulmonary vein which runs from the lung to the left auricle and carries oxygenated blood. The point is that veins always run TOWARDS the heart whereas arteries run AWAY FROM the heart.

The walls of the veins are much less muscular than those of arteries and they have non-return valves (*i.e.* which permit no back-flow of blood) along their length. The pressure of blood is low in the veins and its return to the heart is assisted by muscular movements of the body which squeeze the veins and thus force the blood along the only direction in which it can move, *i.e.* back to the heart.

8. The supply of oxygen to the tissues. Having passed across the thin wall of the lung capillaries the oxygen is transported via the blood to the muscles and other body tissues where it is used in respiration. The transport of oxygen takes place in a number of stages:

(*a*) *The importance of haemoglobin.* As already described an oxygen gradient exists across the alveolar walls. On one side it is at the concentration found in the atmosphere while on the other the concentration (*i.e.* in the blood) is much lower.

Oxygen has a low solubility in water and blood plasma is only a dilute solution. Only 1 cc oxygen will dissolve in 100 cc plasma. A special oxygen carrier which has a very high affinity for the gas is found as the haemoglobin of the red corpuscles. This increases the O_2 carrying capacity of blood up to 20 cc/100 cc blood, *i.e.* by a factor of $\times 20$.

(*b*) *Formation of oxy-haemoglobin.* In the presence of oxygen haemoglobin reacts to form oxy-haemoglobin. When fully saturated human haemoglobin can hold four molecules of oxygen. The equation may be represented thus:

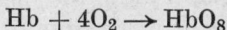

$$Hb + 4O_2 \longrightarrow HbO_8$$

Oxy-haemoglobin is bright red in colour and contrasts with the darker bluish of the haemoglobin itself. Arterial blood thus differs in colour from venous blood, as in the tissues where oxygen is being used the oxy-haemoglobin dissociates, in other words it breaks down to haemoglobin with liberation of the oxygen, *i.e.*

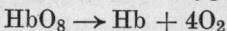

$$HbO_8 \longrightarrow Hb + 4O_2$$

NOTE: Unfortunately haemoglobin has even greater affinity for the gas carbon monoxide (CO) than for oxygen so that if exposed to carbon monoxide it picks it up rather than oxygen. This compound is called *carboxy-haemoglobin* and is stable, so that CO rapidly causes suffocation.

(*c*) *The lymph as a "middle man."* Oxygen released from the oxy-haemoglobin will diffuse across the capillary walls and into the lymph. This is a colourless fluid somewhat similar in composition to blood plasma. It bathes all the tissues of the body directly so that anything passing from or to the bloodstream and the tissues must go via the lymph.

Oxygen passes across the lymph in ordinary solution and thence diffuses into the tissue cells. Here the gas is used in the aerobic stages of respiration by which most ATP is generated.

(d) The route summary for oxygen is thus

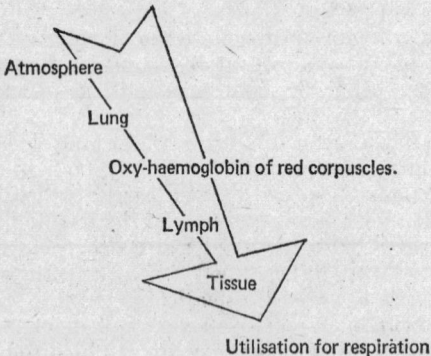

Atmosphere

Lung

Oxy-haemoglobin of red corpuscles.

Lymph

Tissue

Utilisation for respiration

9. Energy exchanges in the tissues. The general importance of the high-energy phosphate, ATP, in energising biological reactions, has been described in 1 of this chapter. As far as the mammal is concerned the most widespread reaction that it powers is that of muscular contraction. For this reason the following refers to muscular tissue although the same general principles apply all over the body.

(a) *ATP generation.* Glucose sugar has been stored in the muscle in the form of the polysaccharide glycogen. This is the major source of oxidation and the formation of high-energy phosphate.

After passing through the intermediate lactic acid (which can be represented as $C_3H_6O_3$, *i.e.* half a glucose molecule) a process which requires no oxygen and only liberates a very small amount of energy, the lactic acid enters the aerobic phase.

This consists of a series of reactions involving many enzymes and taking place within the mitochondria of the tissue. During these reactions lactic acid is broken down to carbon dioxide and water. At the same time the energy contained in the molecule is transferred to make about 36 ATP for each g/molecule of glucose used. The generation of ATP in muscle is therefore:

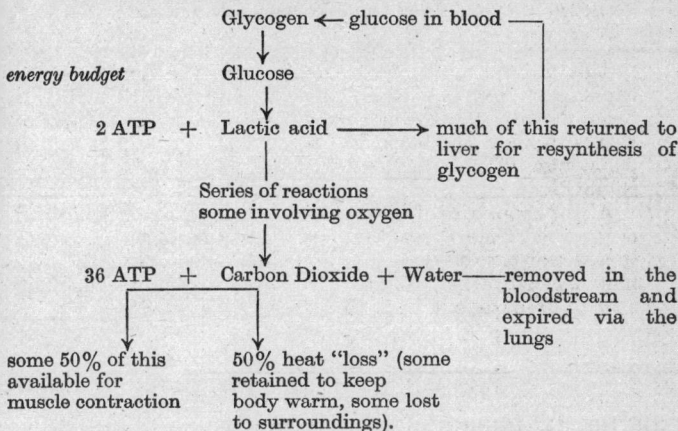

energy budget

Glycogen ← glucose in blood ⎯⎯⎯⎯

Glucose

2 ATP + Lactic acid ⎯⎯⎯⎯→ much of this returned to liver for resynthesis of glycogen

Series of reactions some involving oxygen

36 ATP + Carbon Dioxide + Water⎯⎯removed in the bloodstream and expired via the lungs

some 50% of this available for muscle contraction

50% heat "loss" (some retained to keep body warm, some lost to surroundings).

(b) *The contraction of muscle.* The details of muscle contraction are beyond the scope of this text. Simply, it is brought about by the conjunction of the two muscle proteins actin and myosin to make a new "shorter" molecule. The general effects of this are that the muscle decreases in length, thus pulling on the bones to which it is attached and bringing about movement of the limb or other part of the body.

The energy which allows connection of the two proteins is provided by ATP formed in tissue respiration. After use this substance is down-graded in energy to ADP which in turn must be recharged by further respiration of glucose.

(c) *Disposal of the products of tissue respiration.* The end-products of tissue respiration are, as seen above, energy, carbon dioxide and water.

(i) The energy is partly utilised for contraction and a further useful quantity goes to help maintain the warmth of the body. Where rapid respiration is taking place, heat is being produced too rapidly for retention and much of it will be lost to the atmosphere in radiation and in other ways. The details of the body's thermostatic control are described in Chapter VII but it should be noted that where the body is dehydrated and can no longer sweat (get rid of excess heat), death will rapidly occur. Heat is thus a real waste product of tissue respiration.

(ii) *Carbon dioxide and water.* These can be dealt with together as they combine to make carbonic acid:

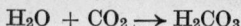

$$H_2O + CO_2 \longrightarrow H_2CO_3$$

This diffuses from the tissue and via the lymph to the capillaries. Here they again dissociate and the carbon dioxide forms a complex with the haemoglobin. The general effect of this is to allow the blood to carry back to the lungs a high percentage of carbonic acid without, at the same time, becoming acid.

At the capillaries of the pulmonary artery the carbonic acid is returned from the red corpuscles to the plasma. Here it breaks down to $CO_2 + H_2O$ and these substances, being in high concentration, diffuse across the alveoli to the lung space. From here it is expired to the atmosphere.

PLANT RESPIRATION

10. General features of plant respiration. Like mammals, green plants also require energy for all the metabolic reactions that maintain life. Unlike mammals, however, plants have a very much lower metabolic rate owing to their lack of movement and low temperatures. A further point of contrast is that plant photosynthesis produces the oxygen which is necessary for aerobic respiration so that during daylight there is a surplus of the latter. This is shown in schematic form in Fig. 23.

(a) *Respiration and photosynthesis.* During photosynthesis oxygen is being produced by the plant but at the same time the gas is being utilised for energy release from synthesised carbohydrates. Energy is required by the plant for building up all its complex organic molecules but on the other hand it has little energy demand for heat and movement.

On balance there is a net gain of stored food to the green plant during light, and a loss during darkness. The former is much in excess of the latter during the summer and the products of photosynthesis are used for storage or for seed and fruit production.

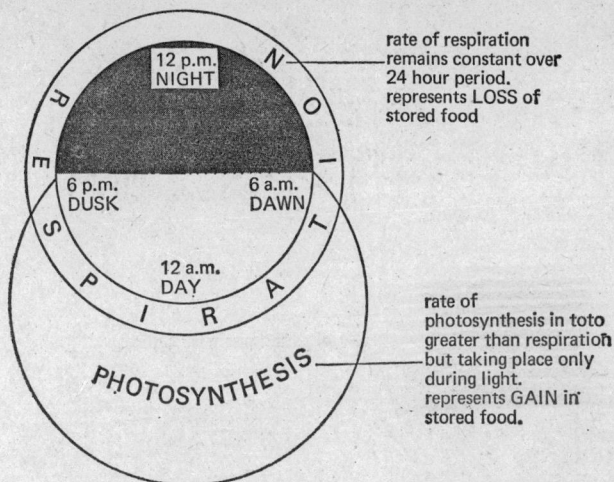

FIG. 23. The relationship between the processes of respiration and photosynthesis in a green plant during a 24-hour period in summer.

(b) *Pathways of oxygen into the plant.* Plants have no ventilation mechanism such as the lung of the mammal. Some oxygen may be taken in by diffusion from the soil and passed up the plant where it is dissolved in the water of the transpiration stream (*see* Fig. 24).

The cuticle of the green parts of the plant and the bark of the older parts are both impermeable to the passage of oxygen but both have a form of pore through which the gas can diffuse. The cuticle is punctuated with stomata which will admit the gas when they are open, and as the bark forms, previous stomatal positions are replaced by lenticels.

Lenticels are eruptions of powdery cork cells in the bark which, while giving protection against water loss, do allow the slow diffusion of oxygen through to the living cambium and parenchyma cells where it is utilised.

11. Factors affecting the rate of plant respiration. Plants exist at the same temperature as the surrounding atmosphere and are therefore much more affected by changes in air temperature than are mammals. The general rule is that for

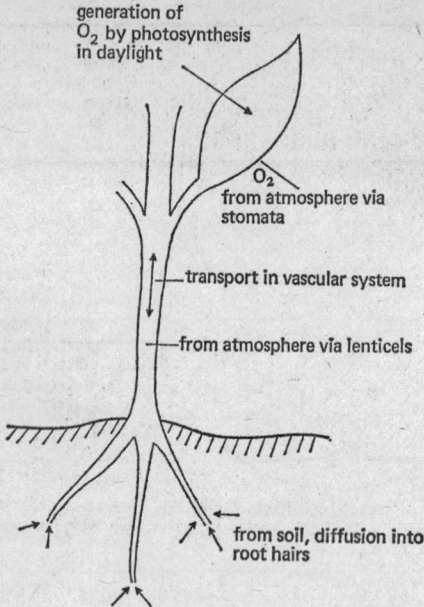

generation of
O_2 by photosynthesis
in daylight

O_2
from atmosphere via
stomata

transport in vascular system

from atmosphere via lenticels

from soil, diffusion into
root hairs

FIG. 24. Sources of oxygen for respiration
by green plants.

every 10°C rise in temperature the rate of respiration will
double until a maximum is reached (about 35°C) whereafter
the plant will suffer destructive chemical changes and
eventually die. Other important factors determining the
respiration rate and products of green plants are as follows:

(a) *Age.* Seeds are dormant stages of the plant and have
a very low rate of respiration enabling them to survive on
stored foods for long periods. Germinating seeds, young
tissues and meristems all have a high rate of respiration
owing to the numerous chemical processes that go on
within their cells. Mature cells as in the normal leaf and
flower have a moderate rate of respiration whereas senescent
leaves and tissues have a low respiration rate.

Thus at any one time the different parts of a plant may
show much greater variations in respiration rates than the
separate organs of a mammal.

(b) *Respiratory material*. As with the mammal the normal respiratory substrate is carbohydrate. The general equation for the respiration of this is:

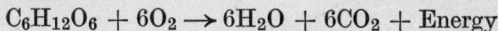

$$C_6H_{12}O_6 + 6O_2 \rightarrow 6H_2O + 6CO_2 + \text{Energy}$$

This means that equal quantities of oxygen are taken up, and carbon dioxide liberated. Some plants, and especially their storage organs (*e.g.* olives, ground and Brazil nuts, cocoa beans, linseed etc.) utilise fat for respiration. Fat is a very economical food store and contains twice as much energy as carbohydrate for the same weight. As fat contains little oxygen in the molecule a lot more is required for its respiration than for that of carbohydrates.

PROGRESS TEST 6

1. What is the importance of ATP? **(1)**
2. Where are the enzymes involved in respiration situated? **(1)**
3. By what means is the pressure of the lung reduced to allow inhalation? **(3)**
4. What are the differences between inhaled and exhaled air? **(4)**
5. Distinguish between an artery, vein and a capillary. **(7)**
6. What is the function of haemoglobin? **(8)**
7. What is lactic acid and how is it formed? **(9)**
8. How does oxygen get into a plant? **(10)**
9. What factors affect the rate of respiration in plants? **(11)**

EXAMINATION QUESTIONS

1. Distinguish between respiration and breathing in man. Devise a simple experiment to find out if, during the anaerobic respiration of living cells, (*a*) oxygen is used, and (*b*) food material is consumed. (*O. & C.*, 1969)

2. Give an illustrated account of the structure and position of the lungs, trachea, and diaphragm of a mammal. (*Cambridge*, 1969)

3. Make a large labelled diagram of the respiratory system of a mammal. Give a brief account of the breathing mechanism in man. (*A.E.B.*, 1970)

CHAPTER VII

THE INTERNAL ENVIRONMENT OF THE MAMMAL

THE CONSTANCY OF THE INTERNAL ENVIRONMENT

1. The necessity for maintenance of a constant internal environment. All the hundreds of reactions of the body's chemistry that are a part of its metabolism are carried out by use of specific enzymes (*see* Chapter II). It is thought that a typical cell might have as many as 1,000 enzymes, and that some of these will be secreted and act outside the cell such as those in the gut and the bloodstream. It will be remembered that enzymes operate in particular conditions but are especially sensitive to the following:

(*a*) *Acidity/alkalinity* (*or pH*). This is the balance between H ions and OH ions. Acid conditions exist where there are more H than OH and alkaline in the reverse situation. The neutral point comes where the balance of opposite ions is equal.

Enzymes are at their most reactive where specific conditions of acidity or alkalinity exist and the correct balance is maintained in the blood and tissues.

(*b*) *Temperature*. While different organisms possess enzymes suited to operate at different temperature optima, it seems that for mammals the balance between maximum rate and the process of breakdown of products is around 40°C. Various means exist for the maintenance of the body temperature and it is common experience that a fluctuation of only a few degrees (as in a fever) causes widespread disorganisation and death.

(*c*) *Degree of hydration*. The immediate cause of death from dehydration is likely to be increase in the viscosity of the blood followed by an explosive heat rise. Enzymes do, however, depend very much on surface reactions and the surface area in dehydrated cytoplasm becomes reduced. Excretion, the elimination of toxic wastes, also depends on plenty of water being available. The body very rapidly adjusts water content to maintain a constant composition in blood and tissues.

(d) *Salt balance.* Many enzymes will act only in the presence of particular mineral ions. The muscles require Ca ions to be present for contraction to take place. If the concentration falls below a certain level the muscles lock in a tetanic spasm.

For the conduction of nerves an exchange of Na and K ions is necessary. The balance of these must be maintained exactly for normal function.

(e) *Elimination of toxic substances.* Enzyme systems may be blocked by the accumulation of the toxic end-products of metabolism. The nitrogen-containing substances such as ammonia and urea, derived in the most part from the de-amination of protein, are particularly harmful. These nitrogenous wastes are kept at very low levels in the body.

Carbon dioxide from respiration would tend to stop the whole series of reactions by which energy is made available. Here again a special region in the medulla of the brain (the region at the top of the spinal cord) is sensitive to the concentration of CO_2 in the blood and causes it to be "blown off" at a rate proportional to the amount accumulated.

It can be seen that for the human body to function normally the internal environment must be maintained within very narrow limits with effect for all the conditions described above. The elimination of waste is called *excretion*.

EXCRETION

2. The wastes, organs of maintenance and their products. The general organs of maintenance and their products are as shown in Fig. 25.

Each of these will be discussed separately in 3–5 below.

3. The lung as an excretory organ. Gaseous exchanges have already been described in Chapter VI. The lung is the major organ for the elimination of CO_2 by respiration in the body. Exhaled air consists of about 4 per cent CO_2. The concentration of CO_2 in the blood is monitored by chemical receptors

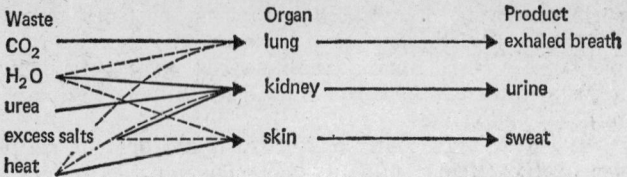

FIG. 25. Waste products and their disposal.

in the medulla of the brain and from this area the ventilation rate is controlled.

Exhaled air is also saturated with water vapour, but in man the lung and tongue play little part in temperature control (unlike the dog) and more water is lost via the kidney and skin. On average some seven joules of heat are lost each twenty-four hours in exhaled air.

4. The structure and functioning of the kidney. The major organs of excretion and internal environment maintenance in mammals are the kidneys. They provide the main means of elimination of nitrogenous wastes, water, excess salts and also of keeping the acid/alkali balance constant.

(a) *Gross structure of the kidney.* If the kidney is cut longitudinally (as shown in Fig. 26) it is seen to consist of an outer cortex and an inner medulla. In the centre is the pelvis of the kidney into which the urine drains. From this leads the ureter which passes down from each kidney to the bladder in which urine is stored. The renal artery enters the kidney from the aorta bringing oxygenated but waste-containing blood. The filtered, but de-oxygenated blood rejoins the vena cava via the renal vein.

(b) *The nephron.* Each kidney contains something over a million individual filtering and re-absorbing units or nephrons (*see* also Fig. 26).

Blood capillaries from the renal artery break up into large numbers of networks called *glomeruli*. Each of these is situated within the eggcup-shaped capsule of the nephron.

From the capsule a long tube leads out, for the most part coiled, but also containing a long single loop—the loop of Henle. A rich blood supply circulates around the tubule, and its capillaries eventually join up into the renal vein.

VII. THE INTERNAL ENVIRONMENT OF THE MAMMAL 91

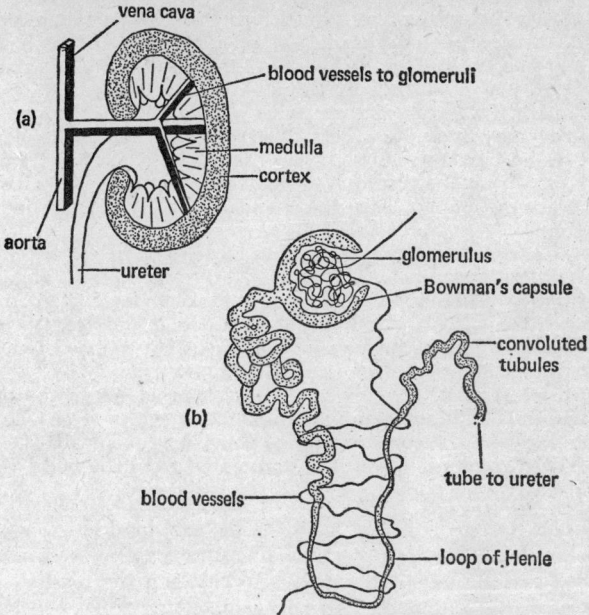

FIG. 26. (a) the kidney, (b) the nephron.

While the capsules are situated in the renal cortex the tubules are found in the medulla and many thousands of them drain into each of the tributaries of the pelvis.

(c) *Functioning of the kidney nephron.*

(i) *Ultra-filtration.* Blood entering the glomerular capillaries is at high pressure and owing to the decreasing bore of the capillaries and their permeability to small molecules a process of "ultra-filtration" takes place. Molecules such as water, glucose, urea, salts, and very occasionally small proteins such as albumin pass out from the blood plasma into the capsule.

(ii) *Selective re-absorption.* All the products of filtration pass from the capsule along the coiled tubule, pushed along by the continuous pressure from behind. During their passage down the tubule useful molecules such as those of glucose are re-absorbed by active processes into the blood capillaries.

Hormones from the pituitary and adrenal bodies control selective re-absorption of mineral salts and the concentration of these actually present in the urine will be dependent on the quantities taken in the diet as well as the needs of the body at any one time. Usually there are more salts in urine than in blood plasma, especially those of Na and SO_4.

(*iii*) *Water content.* As with salts the water content of urine may fluctuate widely. The concentration of the blood is "tested" by receptors in the brain and a special hormone secreted by the pituitary which controls the extent of water re-absorption by the tubule and especially by the loop of Henle.

Under normal conditions some 98 per cent of the water that is filtered into the capsule is re-absorbed as it passes along the tubule. Where a lot of water has been lost from the body by sweating the urine will tend to be strong and scanty. Conversely cold weather when little sweat is produced tends to cause production of more copious dilute urine.

(*iv*) *Acid/alkali balance.* Complex processes in the blood and in the kidney cause mopping up of excess H or OH ions and these are incorporated into buffer ions such as HCO_3 and HPO_3 which are selectively eliminated from the body.

(*d*) *Elimination of urine.* The final product from all the tubules passes via the urethra to the bladder. Here it accumulates until a certain pressure has built up. By reflex action the sphincter muscle between the bladder and the urethra is opened and the urine flows down the latter to the outside. While such an action is entirely reflex in the young human it is obvious that it comes under conscious control later!

5. The skin and temperature control. Besides its importance in temperature regulation the skin also acts as a protective layer and as a sensory region. Only those aspects important in regulation will be described here.

(*a*) *Structure.* Human skin consists of an outer epidermis and an inner and much thicker dermis. Within the latter are the sweat glands and capillaries that are a part of the temperature-regulation mechanism while below it are layers of fatty cells which provide a permanent insulation to the body.

(*i*) *The capillary system.* From the underlying blood vessels a complex and extensive capillary network arises and spreads out immediately below the epidermis. The vessels that supply this network are under sub-conscious control and the volume of blood circulating may vary by as much as a hundred times.

(*ii*) *Sweat glands.* Branches from the capillary network pass into coiled tubular sweat glands. From these, ducts lead up to and open as sweat pores on the surface of the epidermis. The density of these glands varies in different parts of the skin.

(*iii*) *The hairs.* In man the number and extent of the hairs is very much less than for the majority of mammals. Although they do stand on end when the body is cold ("goose-flesh") because of their scarcity they contribute very little to temperature regulation.

(*b*) *Function.* Basically, the skin works in ways which either promote heat loss or encourage its retention.

(*i*) *Heat loss.* To increase heat loss, blood supply to the capillaries near the surface is at a maximum rate and heat is lost to the surroundings by the processes of conduction, convection and radiation. The secretion of sweat which evaporates from the surface of the skin takes a lot of heat from the body. (Evaporation of 1 g sweat requires some 2,000 J.) The hairs of the skin are relaxed so that the minimum of insulation is given.

(*ii*) *Retention of heat.* Here the skin structures work in conjunction with other organs of the body. By decreasing the blood flow in the capillaries of the dermis much less heat is lost. Sweat is no longer secreted. The hairs may be raised, giving rise to "goose pimples," but these are so scanty in man that little insulation results. In other mammals such as cats raising the hairs would greatly increase the warm layer trapped above the epidermis. At the base of the dermis is a layer of fat which retains heat in the deeper tissues by its insulation effects.

While these measures decrease heat loss from the surface of the body extra metabolic heat is generated by involuntary muscular activity or shivering.

(*iii*) *Temperature control centre.* This is situated in the hypothalamus of the brain. It has sensory cells which monitor the temperature of the blood and initiate changes which will lead to correction back to the steady state. This, like respiration, heart beat and gut movements, is another example of an involuntary activity controlled by the central nervous system.

6. The skin as a protective layer. Much of the dermis is composed of tough connective tissue which despite its elasticity protects the body from mechanical damage. It also has strong powers of repair. The Malpighian layer at the base of the epidermis is continually budding off new cells which become keratinised and eventually die and are sloughed off the surface of the skin. These keratinised cells are waterproof and tough and because they are continuously being lost it is not easy for bacteria to get a hold on them. Besides this, the secretions from the glands of the skin may prohibit bacterial growth in the same way as tear drops. Where extra wear takes place as on the soles of the feet and at the joints the Malpighian layer is stimulated to make more layers of epidermis.

MAINTENANCE OF THE BODY AGAINST MICRO-ORGANISMS

7. Pathogenic organisms. While many types of bacteria, fungi and lower animals are quite unable to invade or survive in the body there are a number which can do so. We call these *parasitic* or *pathogenic* organisms.

The body reacts to these in a variety of ways which tend to prevent their gaining access and multiplying within the tissues. Details of the most important classes of pathogens are to be found in Chapter XI.

8. Outer defences of the body. Clearly any pathogen has to enter the body via the skin or else by the invaginations of other systems such as the gut, lungs or urino-genital tubes. Both the skin and the linings of the tracts leading into the deeper systems are covered with a protective epithelium. All epithelia have a high rate of replacement of cells so that it is difficult for pathogens to get a hold on them. Secretions may also be produced which are germicidal. In the trachea and bronchi, mucus entraps foreign material and is then driven back out of the system by ciliary action. The stomach and urinary tracts are acid and inhibit development of micro-organisms.

9. The inner defences of the body. If any of the outer defences described above are breached, say by a cut with a dirty instrument or a thorn, then pathogenic organisms may be introduced into the deeper tissues. If this happens the body has a number of secondary defences. These include:

(a) *High temperature.* The normal temperature of the body (37°C) is too high for the development of many micro-organisms. Local or general rises of temperature upon infection make conditions even more unfavourable to them.

(b) *Antigen–antibody reactions.* Some sorts of white corpuscles and especially those originating from the lymph system have the ability to secrete chemicals which will destroy pathogens in the blood or tissues. The pathogens are called the *antigens* (*i.e.* act against the person) while the chemicals that destroy them are termed *antibodies.* Antibodies act in a variety of ways such as precipitation or solution of the antigens and are quite specific to a particular antigen. Once the body has produced such molecules (say against tuberculosis) it "remembers" how to do so on a second invasion and thus acquires immunity. There are a number of ways of conferring immunity artificially.

(c) *Phagocytic action.* The large multi-shaped white corpuscles in the blood are phagocytic and can engulf and destroy foreign material. In fact they often work after the latter has been inactivated by antibodies. These phagocytes collect at the site of an infection and it is their remains together with dead bacteria that form pus.

(d) *Antitoxins.* Many pathogens produce secretions or waste products that are more harmful to the body than the actual presence of the micro-organism itself. These are called *toxins* and are among the most poisonous materials known.

The secretory white corpuscles that produce the antibody also make chemicals called *antitoxins* which neutralise the toxins of the parasites. Again antitoxins are quite specific to the particular toxin molecules.

SUMMARY OF THE INTERNAL ENVIRONMENT

10. The internal environment. As we have seen from the foregoing sections the mammalian body has a complicated and constant internal environment whose precise constancy must be maintained for life to continue. By the action of a number

of tissue and organ systems the body succeeds in the maintenance of a very exact chemical composition, temperature,
degree of hydration and freedom from parasitic invasion.
Should any of these controlling mechanisms break down, for
one reason or another, then death will be the result.

It should be noted that the constancy of the internal
environment is very much more exactly maintained in higher
animals such as the mammals than in lower animals (such as
worms). Its exact regulation in all sorts of external environments is one of the major reasons underlying the success of
the mammals as a class.

PROGRESS TEST 7

1. What factors in the body are kept constant? (1)
2. List the excretory products of the mammal. (2)
3. Of what does the nephron consist? (4)
4. How is the water content of urine controlled? (4)
5. What parts of the skin are important in temperature
   control? (5)
6. How does the body protect itself against the entry of
   bacteria? (8)
7. What are antibodies and antigens? (9)
8. What role do phagocytes play in the defence of the body
   against infection? (9)

EXAMINATION QUESTIONS

1. What is the normal temperature of man? Make a list to show
the ways in which heat may be gained and lost from the human
body. Which of these may be controlled? Describe how man
conserves his body heat. (O. & C., 1972)

2. Explain what is meant by the term "metabolic wastes."
Explain the role of the kidney in
    (a) elimination of toxic products from the body
    (b) maintenance of a stable internal environment.
                                        (O. & C., 1970)

3. By means of labelled diagrams and explanatory notes indicate
the structure of the mammalian kidney with its blood supply.
What is urine and how is it formed? (A.E.B., 1970)

4. Describe the events taking place at the point of the body that
has been cut with a dirty razor blade. (after O. & C.)

CHAPTER VIII

THE CYCLING OF NUTRIENTS IN NATURE

CARBON CYCLE

1. Cycling of nutrients as an interaction between plants, animals and their environments. We have now covered sufficient of the basis of plant and animal nutrition and of the raw materials and excretory products of living organisms to see the interconnections that exist. Because CO_2 is respired by all living things and yet taken up and fixed by plants which in turn are eaten and respired by animals it may be said to *cycle* in nature. This is also true for nitrogen and indeed for every single element that is passed through food chains from one level to another.

An important link in all cycles in nature are the saprophytic organisms that exist in the soil and on the sea bed and cause the decay of the dead bodies and excretory matter of higher organisms, and which return their constituents into the medium.

The discovery of the details of nutrient cycles is scarcely a century old and the gradual growth of understanding of their nature has helped us understand how greater productivity can be obtained by assisting the natural processes.

Because of their overall importance the carbon and the nitrogen cycle will be considered in greater detail (*see* also Chapter III on the plant and the soil).

2. The carbon cycle.

The carbon cycle has been mentioned in the chapter on plant nutrition, and its gains and losses have been dealt with in the parts of the text that relate to photosynthesis and respiration. It can be seen that the straightforward uptake and loss of CO_2 by photosynthesis and respiration is made more complex by the cycling of much organic matter through death and decay. By no means all of the CO_2 is returned to the air or water through decay, as under anaerobic conditions carbon compounds accumulate as fossil fuels. In recent years there has been a steady but small increase in the percentage of CO_2 in the atmosphere through industrial consumption of fossil fuels. Increased CO_2 should lead to higher plant yields but the associated "smog" and pollution that goes with industrial activity cuts out a part of the sun's light and may also insulate us from the solar heat. The long-term effects of increasing the CO_2 content of the atmosphere are somewhat controversial. Figure 27 shows the carbon cycle diagrammatically.

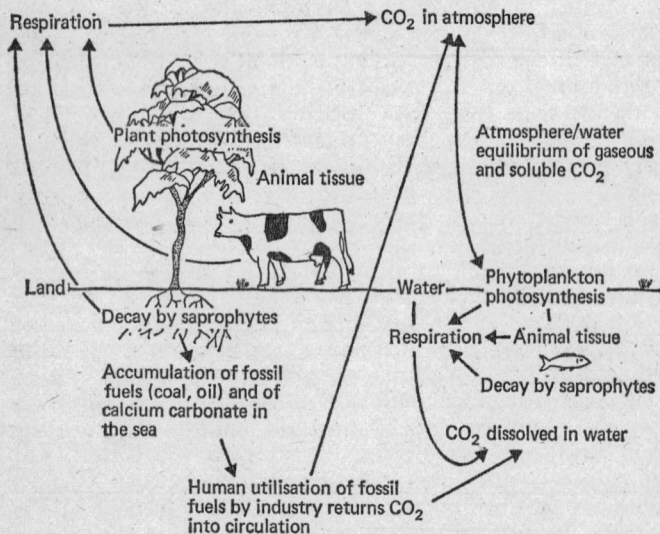

FIG. 27. The carbon cycle.

3. **Some figures relating to the carbon cycle.** It should be appreciated that these figures are only estimates and different authorities quote different ones although they do approximate to the same order.

It is at least of interest to examine the earth's annual carbon budget and reserves and to note how very little of this actually becomes available to man partly because of its unavailability (most sea productivity and partly through losses in food chains (*see* **8**)).

Quantity of CO_2 fixed each year by plants $\rightleftharpoons 20 \times 10^{12}$ kg
Quantity of animal tissue this can support $\rightleftharpoons 20 \times 10^{10}$ kg
Total return of CO_2 to the atmosphere by respiration
$\rightleftharpoons 20 \times 10^{12}$ kg
Amount of CO_2 present in the atmosphere $\rightleftharpoons 800 \times 10^{12}$ kg
Amount of fossil fuel laid down in each year $\rightleftharpoons 1 \times 10^7$ kg
Annual rate of consumption of fossil fuel $\rightleftharpoons 2 \times 10^{12}$ kg
Reserve of fossil fuel $\rightleftharpoons 10,000 \times 10^{12}$ kg
Reserve of CO_2 locked up as limestone in rocks
$\rightleftharpoons 50,000,000 \times 10^{12}$ kg

In the final sections of this chapter some of the more obvious implications of the studies of these cycles to the human situation will be considered.

NITROGEN CYCLE

4. The nitrogen cycle. Most of the nitrogen in the atmosphere is quite unavailable to plants. The available nitrogen in circulation is in the form of nitrogen salts or organic molecules and is a strictly limiting factor in plant productivity.

Nitrogen is present in all amino acids and proteins and nucleic acids, thus the circulation of this element in nature is second only to carbon in its importance. Unlike the information derived from the carbon cycle a great deal of direct application to increased yields has been possible by understanding nitrogen circulation, and the really scientific basis of increasing crop productivity can be related to its exposition in the 1870s.

For our purposes the cycle will be depicted on land but it should be clear that a similar sort of cycling takes place in the sea and in fresh water with all the substances being in solution. The cycle is summarised in Fig. 28.

5. Applications of the nitrogen cycle. These may be considered in terms of increasing gains and reducing losses.

(a) Increasing gains. These put more nitrogen into the soil and thus produce a higher yield of crops. In nature only four units of nitrate are actually turned into plant protein on an acre in a year. The modern farmer expects to produce a figure around twenty-five and as much as sixty has sometimes been achieved.

Nitrogen gas in atmosphere

Plant protein

→ Animal protein

Fertiliser industry

→ Excreta

Soil

Uptake of nitrates by roots

Death and decay

Nitrifying bacteria free in soil or on plants roots of Pea family fix nitrogen to nitrate

Saprophytes turn all protein to ammonium (NH_3) compounds

Nitrosomonas type bacteria convert ammonia to nitrite ($NH_3 \rightarrow NO_2$)

Denitrifying bacteria work in anaerobic soils convert nitrate back to nitrogen gas (N_2)

Nitrobacter type bacteria convert nitrite to nitrate ($NO_2 \longrightarrow NO_3$)

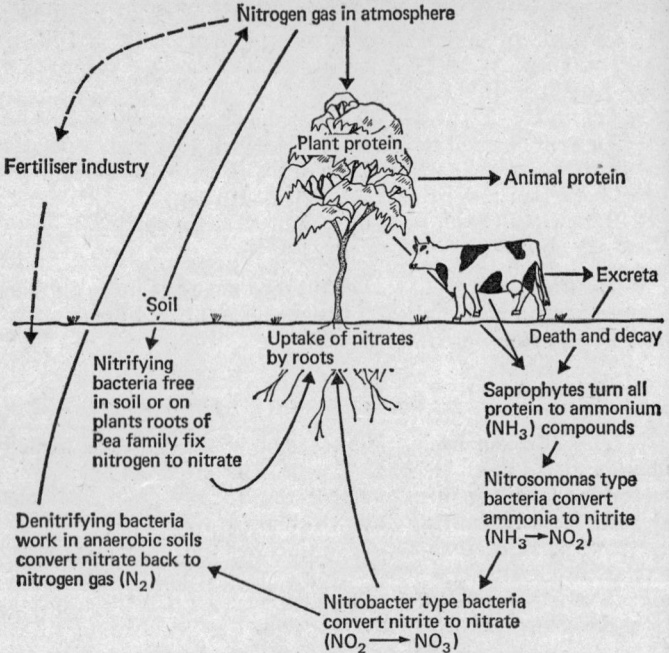

FIG. 28. The nitrogen cycle.

(i) *Aerobic conditions.* Those bacteria which break down excreta and dead tissues to produce nitrate work only in aerobic conditions. For this reason draining and aerating the soil is most beneficial to the natural operation of the cycle. Liming also produces favourable pH for their activities.

(ii) *The addition of extra nitrogen* to the soil in the form of natural fertilisers such as manure, or artificial ones such as ammonium salts and nitrates, adds greatly to the availability of nitrogen to the plant roots. As nitrate is a limiting factor to plant growth, yields may be in direct proportion to extra nitrogen added.

(iii) *Use of leguminous crops.* Members of the pea family such as beans and clover develop nodules on their roots. Within these nodules the Rhizobium bacteria fix atmospheric nitrogen in the soil and turn it direct to nitrate for plant use. The alternating of leguminous with other crops and the subsequent ploughing in of the legume vastly increases the store of nitrogen in the soil.

(b) *Minimising losses.* In a sense the positive steps taken to increase gains will also minimise losses. There are however certain specific activities which recommend themselves. These are:

(*i*) Improvement of the tilth and composition of the soil towards a fertile humus-containing loam will discourage the activities of denitrifying bacteria which reduce fertility in sour anaerobic soils.

(*ii*) Application of fertilisers in a form where the extra nutrient is actually taken up by the soil and not allowed to run away in soluble form.

(*iii*) Recycling of organic matter such as sewage back on to the land where it can be reutilised. This is far better than its disposal into fresh water where it causes loss of oxygen, and thus deterioration or disposal into the sea where it becomes lost from the productive cycle.

6. Other cycles. It will be recalled that plant growth is limited by other elements besides nitrogen and in particular potassium, sulphur, phosphorus and calcium have been shown to be important. All of these, as well as the many other trace elements, have their cycles in the soil. The major nutrients are now all supplied to the soil in the form of fertilisers in quantities which preclude their acting as limiting factors.

By manufacture and application of fertilisers the productivity per acre in developed countries may be as much as fifteen times greater than unassisted nature could be expected to do due to the speed at which these various cycles turn over in the soil.

The increase of productivity by the development of fertilisers is a part of the agricultural revolution which has included the application of mechanisation to farming and of genetics to the improvement of crop yield. Development of systemic pesticides and herbicides has been another important field of applied biology.

ENERGY RELATIONS

7. Energy flow through food chains. Examination of any food chain of the plant–herbivore–carnivore type shows surprisingly high losses of energy between stages. The plants which use the energy from the sun are called the *primary producers*. In fact they only manage to capture less than one per cent of this energy and they utilise this for their own growth and metabolism. Some of the tissues they have made will in turn be passed on to herbivores who also use their food for growth but also for their own metabolism. By the time we come to the carnivore level we may have only $\frac{1}{500}$ or so of the original fixed plant material retained.

It is very important to understand this fact in terms of taping a food chain at different levels. Clearly it is much more efficient to grow vegetable crops in terms of yield than for us to feed on herbivores and, except for fish, it would be quite impractical to try to live off carnivores.

On the other hand vegetable protein is less useful and concentrated than animal protein and certainly less palatable as a standard diet. It may be that the next few decades will see a compromise, with the developed nations reducing to a less extravagant animal diet, and the developing nations being able to bring up the protein levels of their own intake. Such ideals are not necessarily easy as political solutions but to biologists considering the situation of our species as a whole they seem absolutely essential goals.

8. Sample food chains showing energy flow.

(a) *Grass to man.*

SUNLIGHT⟶GRASS   energy utilised: 6,000,000 units
energy respired: 5,000,000 units
energy passed on to herbivores such as cows or sheep: 1,000,000 units

HERBIVORE   energy utilised: 1,000,000 units
energy respired: 50,000 units
energy passed on to carnivore such as man: 50,000 units

CARNIVORE—the dominant member of this food chain percentage of original plant growth represented in human diet

$$= \frac{50,000}{6,000,000} \times 100 = 0.83\%$$

actual human tissue increase $\simeq 0.14\%$

(b) *Trees in a wood to a fox.*

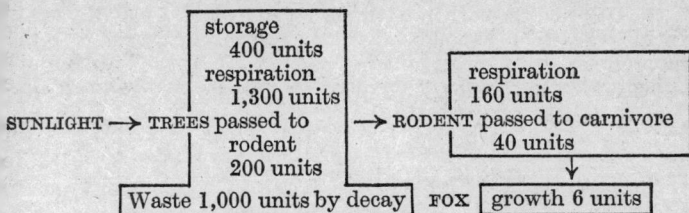

```
                    ┌─────────────────┐
                    │ storage         │
                    │ 400 units       │         ┌────────────────────────┐
                    │ respiration     │         │ respiration            │
                    │ 1,300 units     │         │ 160 units              │
SUNLIGHT ──→ TREES  │ passed to    ──→│ RODENT  │ passed to carnivore    │
                    │ rodent          │         │ 40 units               │
                    │ 200 units       │         └────────────────────────┘
                    └─────────────────┘                    ↓
        ┌─────────────────────────────┐         ┌────────────────────────┐
        │ Waste 1,000 units by decay  │  FOX    │ growth 6 units         │
        └─────────────────────────────┘         └────────────────────────┘
```

percentage of original plant energy $\dfrac{6}{2900} \times 100 = 0.21$

From these two examples the nature and extent of the energy losses involved in food chains can be seen. Any other examples would produce much the same result and the longer the chain the more drastic are the losses.

PROGRESS TEST 8

1. How is carbon taken up from the atmosphere? (2)
2. What are fossil fuels? (2)
3. Why is the CO_2 content of the atmosphere increasing? (3)
4. What role do micro-organisms play in the nitrogen cycle? (4)
5. What is the importance of nitrogen fixing bacteria? (5)
6. How may losses of nitrogen from the soil be minimised? (5)
7. Why does a herbivore not pass all the energy it consumes to a carnivore that eats it? (8)

EXAMINATION QUESTIONS

1. Describe the carbon of the nitrogen cycles. (*A.E.B.*, 1970)

2. Use each of the following words once only and write it on the appropriate dotted line against an arrow or in a box on the diagram. Energy; Decay; Carbon Dioxide; Respiration; Herbivore; Plant; Carnivore. (*after O. & C.*)

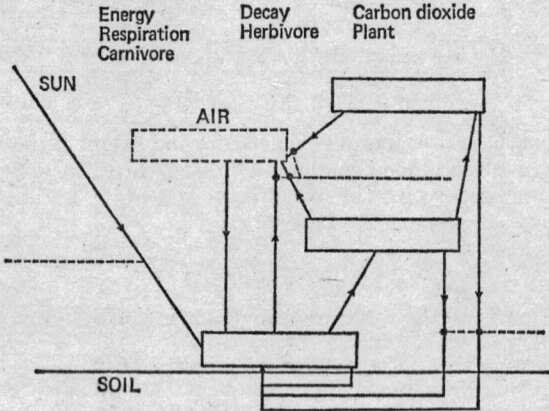

CHAPTER IX

THE CO-ORDINATION AND MOVEMENT OF PLANTS AND MAMMALS

STIMULI AND THEIR RECEPTION

1. The nature of stimuli. Animals and plants are able to detect and respond to many sorts of changes in their environments—both external ones as well as those within their own bodies. The changes that bring about a response are those related to the survival of the organism, thus living things do not respond to radio waves but most of them do respond to light as a stimulus.

The elaboration of stimulus detection and response is much greater in animals than plants but both groups do in fact respond to much the same types of general stimuli.

(a) *Gravity* is an important stimulus to animals as it allows them to maintain balance and adjust to changes in posture involved in locomotion. The basis of gravity detection may be some sort of "stone" of heavy material on a bed of nerve cells or it may be by the movement of a fluid pulling at nerve cells, both these factors being present in the human ear. Within the tissues and muscles gravity detection is done by the stretching of nerve cells that are in parallel with tissue cells.

In plants gravity is important to indicate a direction for the primary growth of shoots and roots on germination and during subsequent development. Roots, which are to provide support for the plant, grow towards gravity (called *positive geotropism*) while shoots, which are to spread out the photosynthetic regions in the air, grow away from gravity (*negative geotropism*).

(b) *Light*. For animals analysis of light stimuli allows recognition of objects in the environment such as food, a mate, predators, etc. For the mammal such as man the wavelength of light can be detected allowing colour vision

and the distance of an object can also be deduced. Special
sensory cells in the retina of the eye change the light that
falls on them into electrical energy which is transmitted to
the brain as nerve impulses.

Light is necessary to all green plants in order that photosynthesis can occur and shoots tend to grow strongly
towards the light (called *positive phototropism*). Roots, on
the other hand, grow away from light and thus into the
ground (*negative phototropism*).

Both plants and animals react in complex manner to
long-term effects of light which effect their reproductive
patterns. On the whole increase of light in the spring
switches on chemical systems that lead to production of offspring at favourable times of year.

(*c*) *Chemical changes.* Gradients of chemicals associated
with the presence and recognition of food or a mate or of
other significant objects in the environment are detected by
animals. The senses of smell and taste are very highly
developed in many mammals although rather less so in man
with his arboreal ancestry.

Whole plants do not respond to chemical gradients
although the sex cells of both plants and animals often find
their partner for fertilisation by swimming or growing down
a chemical gradient.

(*d*) *Mechanical changes.* For animals these include the
sense of touch which gives them immediate information
about the environment with which they are in contact, and
their sense of hearing which gives information about changes
taking place at a distance.

Many plants do not respond to touch at all, but climbing
plants such as convolvulus grow unequally when in contact
with an object and thus come to curl around it. A few
plants such as the mimosa and Venus fly trap respond very
rapidly to touch.

(*e*) *Other stimuli.* Plants and animals respond generally to
heat by more rapid growth, but also in specific ways. An
excess of heat, as with any other stimuli, is interpreted by
the body as pain and this feeling, despite its unpleasant
nature, is an important means of ensuring survival for
animals. Plants, of course, do not feel pain although they
may react to mechanical damage caused of wounding by new
growth.

Oat coleoptile placed
in unidirectional light

light

After 2 hours the tip of
the plant has turned towards
the light. This is positive
phototropism

If the tip of the coleoptile is removed
or covered

No change in the direction of growth
takes place. This shows that the
region sensitive to directional light
is the tip of the shoot.

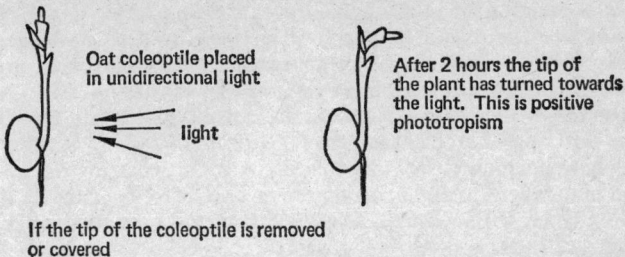

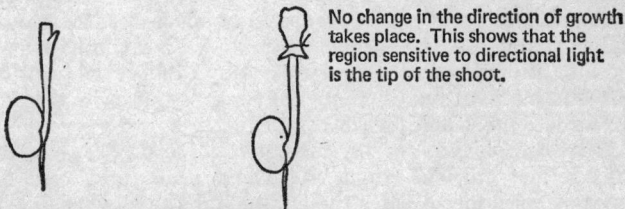

If a coleoptile has its tip removed then
a block of Agar jelly placed between the tip
and the rest of the shoot and again placed
in unidirectional light it will respond

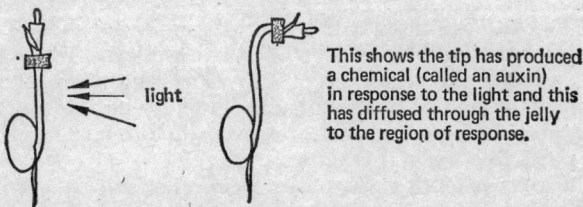

light

This shows the tip has produced
a chemical (called an auxin)
in response to the light and this
has diffused through the jelly
to the region of response.

A clinostat is an instrument for revolving
a plant four times an hour. If this is done then
both root and shoot grow horizontally — if the
rotation is stopped the shoot turns upwards and the
root downwards (positive geotropism).

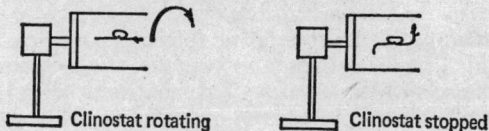

Clinostat rotating

Clinostat stopped

Fig. 29. Tropisms in plants.

2. Detection of stimuli and response in plants. Most of a plant is either dead in the form of wood or else its cells are fully differentiated and neither react nor respond to stimuli. At the apical meristems however, that is, at the shoot and root tips, the situation is different. Here cells which are dividing but not yet differentiating produce auxins or plant hormones.

These hormones are translocated back from the meristems to the elongating regions behind them and here they materially affect the rate of cell expansion and thus the rate and direction of growth.

The receptor to light, gravity or touch is thus in the apical meristem and the region of response, or effector, is the growing region behind it. The responses are very slow compared with those of animals and they take place by differential growth.

A summary of experiments showing response in plants to uni-directional stimuli is shown in Fig. 29.

More complex responses, for example to day length, take place in leaf and bud primordia. There are a number of other types of hormones besides the auxins but the whole subject is extremely complicated and by no means fully understood.

SENSE ORGANS OF MAMMALS

3. The eye. For man, rather more than 70 per cent of all incoming information is received through the eyes, which are the most important of our receptor systems. As explained in **1** it is the function of the eye to turn light into nerve impulses which can be relayed and interpreted by the brain. The eye is made up as follows, and is shown in Fig. 30.

(a) *The conjunctiva* is the outer covering of the eye. It is a thin layer of living cells and has no particular significance in vision.

(b) *The cornea* is the front part of the sclera, which is the tough fibrous coat that surrounds the eye. It is transparent and because of its refractive index it bends light rays inwards to help focus them on the back of the eye.

(c) *The iris and pupil.* Light entering the cornea now passes through the pupil, which is a hole surrounded by a coloured diaphragm of muscle called the *iris*. This contracts according to the amount of light falling on the eye and cuts down the amount to give maximum acuity. In this respect its function is exactly analogous to the stop on a camera.

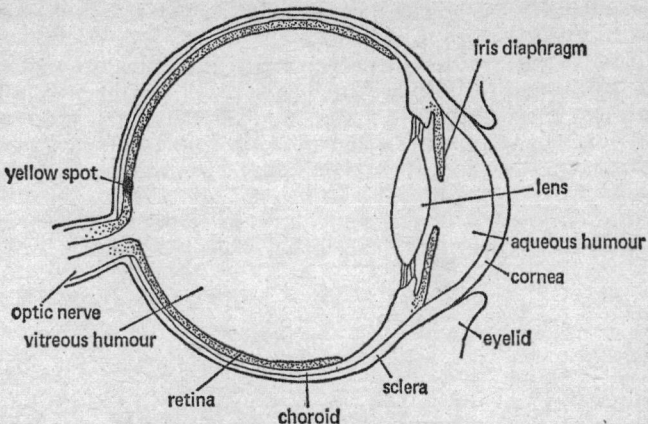

FIG. 30. The eye seen in sagittal section.

(d) *Aqueous humour*. This is a transparent fluid between the lens and the cornea.

(e) *The lens and ciliary apparatus*. The lens is elastic and re-fracts light differently according to its shape. This in turn is changed by the ciliary muscles and ligaments. The process of changing the lens shape is called focusing and it leads to an image being formed exactly on the sensory retina at the back of the eye.

In normal circumstances the lens is stretched out flat so that it has a long focal length suited to looking at distant objects. When one requires to observe near objects the ciliary muscles contract and the lens bulges, decreasing its focal length. It is thus more of a strain to observe near than distant objects.

(f) *The vitreous humour*. This is more viscous than the aqueous humour although it is also quite transparent. It supports the inside of the eye and at the same time allows light to pass through it and fall on the retina.

(g) *The retina*. This consists of many millions of sensory cells which have the ability to convert light into nerve impulses. There are two sorts of receptor cells, the rods which are only sensitive to the intensity of light (and so respond only to black and white) and the cones which react to different wavelengths and thus allow colour vision. These latter are most densely congregated at the yellow spot which is the part of the eye on which light is focused.

Each sensory cell has a nerve process and these are between the retina and the vitreous humour so that the light must pass through the nerve fibres. All the fibres collect together and leave the eye at the blind spot (so called because light falling on it cannot be seen). The nerve running from the retina to the back of the cerebrum is called the *optic nerve*.

(*h*) *The choroid*. A pigmented layer underlies the retina and prevents internal reflection of light. This is called the *choroid*

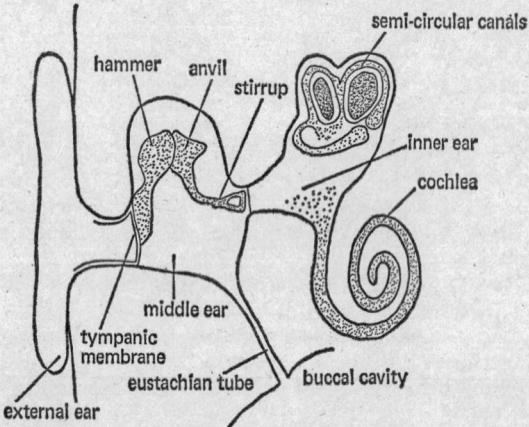

FIG. 31. The ear in frontal section.

and it is this black central area we see as we look through the pupils of our eyes in a mirror.

(*i*) *The sclera* is the elastic coat which surrounds and protects the whole of the eye. To it are attached the eye muscles which allow rotation in the socket. The two eyes overlap much the same fields of vision and give us the ability to judge distance or stereoscopic vision.

4. The ear (*see* Fig. 31). Sound reception is important to all mammals but in the case of man and his development of language it has come to have a special significance. The ear not only tells us the direction and meaning of sounds but is also the main organ of balance of the body.

(a) *The outer ear.* The pinna collects up sounds in the same way as an ear trumpet and directs them down the external auditory meatus to the ear drum. The ear drum is a tightly stretched membrane which picks up vibrations of the air and passes them to the middle ear.

(b) *The middle ear.* The chamber known as the middle ear connects to the back of the mouth via the Eustachian tube. This allows the equalisation of pressures on either side of the tympanic membrane. Sounds in the form of vibrations are conducted across the bones of the middle ear called the *hammer* (malleus), *anvil* (incus) and *stirrup* (stapes). These act as a series of levers which magnify the small movements of the tympanic membrane, passing it to the oval window of the inner ear. This window is $\frac{1}{22}$ the size of the tympanic membrane so that the force acting on it is 22 times greater.

(c) *The inner ear* is where sound vibrations are converted into nerve impulses and where the organs of balance are found. It is best considered in terms of these two functions.

(i) *The cochlea* is the lower part of the inner ear and consists of a spiral tube across whose section are stretched two membranes. The whole system is full of fluid. Within the membranes nerve cells are situated and tiny movements of the membranes, passed through the fluid, cause nerve impulses to be generated by the nerve cells.

From the sensory cells the auditory nerve communicates with the side portions of the cerebrum responsible for sound discrimination.

(ii) *The balance centre* is in the upper part of the inner ear and consists of three semi-circular canals in which movements of fluid against nerve cells indicate any turning positions to the brain. Besides this there are three ear stones which lie on beds of nerve cells and movement of the body in linear (straight line) directions are detected by these.

Both types of apparatus help us to maintain our balance and communicate with the brain via the auditory nerve. In this case however the nerve fibres enter the cerebellum which is the part of the brain concerned with balance and locomotion.

5. The nose is the organ responsible for the detection of smells and there are hundreds of different smells that we can distinguish. It is not very clear exactly how the nose works but it appears that some molecules have the property of stimulating the olfactory epithelium (sensory area within the nasal region) to produce nerve impulses. These are transmitted to a region in the front of the cerebrum by the olfactory nerves.

6. The tongue is responsible for the sense of taste. There are in fact only four types of taste and the buds on the tongue are distributed differently in respect of these tastes. Thus the tip and sides of the tongue are sensitive to sweet and sour and the back to bitter and salt tastes. It will be noted that whereas we can smell things at a distance we can only taste them when they are in contact with our tongues and this is another distinction between these two related senses.

7. The skin has already been dealt with as a protective organ and for its role in the maintenance of temperature. Within the dermis however there are also receptors for touch and for the detection of heat and cold. Rather diffuse nerve endings seem to be responsible for the detection of pain as a stimulus.

NERVOUS CO-ORDINATION IN MAMMALS

8. Nerve cells are extremely specialised. They have at the near end a system of connections called *synapses*. These run into a larger conducting portion called the *dendron* which in turn leads to the cell body. From the cell body the axon leads out and this finally breaks down into a number of synapses or end plates.

Intermediate cells and brain cells have axons and dendrons of equal length, while sensory cells have extremely long dendrons and motor cells long axons.

Nerve impulses are conducted at a rate of some 100 m/sec and consist of an electric current of some 100 mV. Different signals run in the nervous system by means of the frequency (rather than the amplitude) of impulses. Nerve cells are usually covered in their conducting regions with a fatty myelin sheath interrupted along its length by nodes.

9. The spinal cord. Sensory information, *e.g.* from the skin receptors, passes by sensory nerves into the top of the spinal cord. Here the sensory nerves communicate or synapse with other nerves giving routes leading up and down the spine as well as those leading directly out to muscle systems.

Outgoing impulses pass down the motor nerves to effector organs which will actually cause something to happen. We thus have the simplest pattern of behaviour in the reflex arc whose sequence is

> stimulus
> receptor
> sensory nerve
> intermediate nerve
> motor nerve
> effector organ (*e.g.* muscle)
> response

Various skills such as walking may be considered to fall in this category, but basically the spinal reflex is protective and allows of immediate action by the body even before the brain has had time to come into play and work out complex solutions. Thus the removal of one's hand from a painful stimulus is an example of a spinal reflex.

10. The brain (*see* Fig. 32). The brain may be likened to a massive extension of the spinal cord. Like the latter it has incoming and outgoing signals but it differs in the great complexity of the intermediate pathways.

(*a*) *Incoming traffic.* Sensory nerves pass either directly or via the spinal cord to the brain from the main sense organs in the head as well as the extensive receptor network in the skin.

(*b*) *Regions of the brain.* The brain has a number of functional regions which include sensory and motor areas as well as memory banks in which information is stored.

(*i*) *The medulla* is at the back of the brain at the top of the spinal cord. It is the centre of much of the unconscious activity of the body such as respiration and heart beat. Tracts also pass through the medulla to other regions.

(*ii*) *The cerebellum* lies immediately above the medulla and is the centre for the co-ordination of balance and posture and locomotion. It receives impulses from the inner ear and from the muscles and sends out motor impulses to the muscles.

(*iii*) *The cerebrum* is the large front portion of the brain and it has a number of different areas. At the back is the optic

The reflex arc

(a)

The brain & central nervous system

(b)

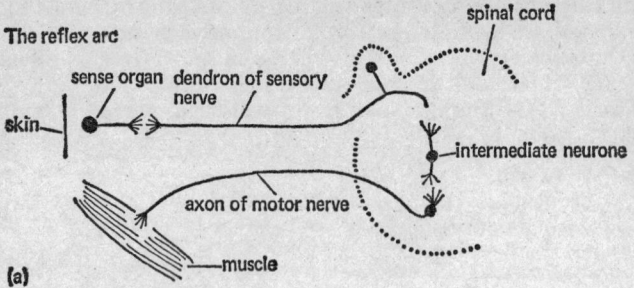

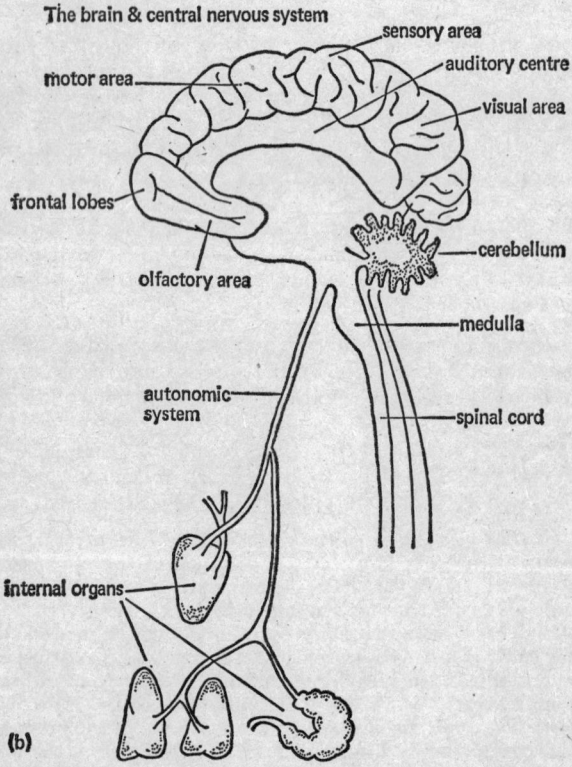

FIG. 32. (a) the spinal reflex, (b) the brain and spinal cord.

region where sensory data from the eyes is interpreted. At the sides are the auditory areas for discrimination of sound while in the middle region is the area for perception of touch. In front of this is the main motor region of the brain which sends motor impulses down the spinal cord to all the voluntary muscles of the body.

In the very front of the cerebrum is the "silent area" which contains all the memory banks of past actions and is the seat of intelligence and behaviour. The olfactory area is small and lies below the frontal lobes.

HORMONAL CO-ORDINATION IN MAMMALS

11. The hormones. Besides the nervous system already described, the body also makes use of chemical co-ordination by means of hormones. On the whole these are slow acting and bring about widespread changes such as maturation, sexual cycles, etc. The emergency hormone adrenalin is an exception to this however, as it acts instantaneously. A summary of the major hormones of the body and their activity is given below.

Name of gland	Name of hormone	Action of hormone
PITUITARY	Pituitary hormones	Promotes growth and maturation; is the master hormone acting on all other hormone glands. If present in excess causes giantism, while a deficiency produces a dwarf.
THYROID	Thyroxine	Controls the metabolic rate of the body. Also important during maturation. Deficiency causes cretinism. In excess causes increase in heart beat, temperature, etc., while if deficient causes opposite effects. This hormone contains iodine and the symptoms of deficiency are sometimes brought about by inadequate diet.
PANCREAS	Pancreatic hormones	Control the sugar level of the blood. Diabetes is the disease when defective insulin is produced and blood sugar levels fluctuate widely.

Name of gland	Name of hormone	Action of hormone
ADRENALS	Adrenalin	The "fight or flight" hormone; it is produced under conditions of stress or anger. It stimulates the body to more efficient performance, increasing heart beat and blood flow to the muscles.
TESTES (of male)	Testosterone	Controls maturation and promotes sperm formation. Provides basis of sex drive and the development of the secondary sexual characters.
OVARY (of female)	Oestrogen	Causes ovulation to occur. It causes maturation, development of the secondary sexual characters and sex drive.
	Progesterone	Produced also by the placenta during pregnancy it causes the female body to accept and develop the foetus, it also stimulates breast development and promotes parental care by the female towards her offspring.

EFFECTOR SYSTEMS IN MAMMALS

12. The muscles. These are by far the most important effector organs of mammals. Muscles consist of specialised contractile cells that are served with motor nerves from the central nervous system. When the brain "wishes" to carry out a specific action, impulses pass from the motor cortex of the cerebrum via motor tracts in the medulla and spinal cord to motor nerves opposite the appropriate muscle system. Impulses travelling in these nerves to the muscles cause them to contract by some 20 per cent of their length. The more nerves and the greater frequency of impulses the more powerful is the contraction of the muscle concerned. In order to understand how muscles actually bring about movement of part of the body several concepts must be understood:

(a) *Muscles and bones.* A given muscle will have an origin

on one bone and its insertion on another, the muscle passing
across a joint. Thus in Fig. 33 the biceps has its origin in the
scapula and humerus and insertion on the radius. The bone
is rigid but at its ends it has an apparatus of cartilage and a

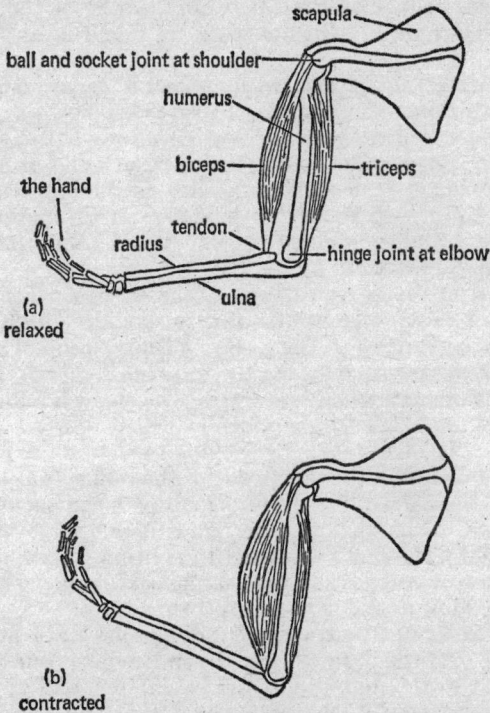

FIG. 33. The working of the forearms: (a) shows the
antagonistic muscles, (b) illustrates the forearm
contracted—the biceps contracts and pulls up the
radius while the triceps is passively stretched.

fluid-filled sac which allows articulation of one bone with
another. The joints are held together by strong but flexible
tissues called *ligaments* while the muscles are attached to
bones by a rather similar tissue called a *tendon*.

(b) *Antagonism in muscle systems*. The only power action

of a muscle is contraction and once contracted the muscle has no ability to expand. For this reason muscles are arranged in pairs with the contraction of one of the pair leading to the passive elongation of the other, and vice versa. As the pair will run on each side of a joint one will cause movement of the distal bone in one direction and the other in the opposite way. Thus we say that muscles are found in antagonistic pairs.

(c) *Movement of the human forearm as an example of muscle action.* The human forearm is as shown in Fig. 33. The upper bone, the humerus, fits into a ball and socket joint at the shoulder girdle. In the front, and attached by two tendons to the shoulder girdle and the humerus, is the biceps muscle. This has its insertion on the radius of the lower arm and runs across the elbow joint. (*See* arrangement of biceps and triceps in Fig. 33.)

Antagonistic to the biceps is the triceps which has an origin of three tendons attached to shoulder girdle and the top of the humerus. The insertion is to the end of the ulna. When the biceps contracts the forearm is raised and when the triceps contracts it is lowered. The elbow is a hinge joint allowing movement in only one plane.

This type of system is repeated in the lower arm, so that a set of muscles on the upper side of the radius lead to extension of the hand while the antagonists on the bottom of the ulna lead to its flexion. The radius is free to roll over the ulna and this allows the hand to turn on its axis. Special provision of muscles and the anatomical arrangements of the thumb allow grasping to take place.

The whole of the arrangement of muscles and bones in the human forearm can clearly be related to our arboreal ancestry as the limb is adapted to climbing and grasping.

(d) *Leverage.* It has been noted that the muscles only contract by about one-fifth of their length and yet very extensive movements can be made. This is done by the leverage operating at the joints. The body normally employs levers that have low mechanical advantage but high velocity ratios. In the ankle, third-order levers with mechanical advantage are employed.

13. Involuntary muscle and the autonomic nervous system.

The type of muscle described in the previous sections has been the voluntary (skeletal) muscle so called because it is under the control of the conscious will. Its contractions are rapid and bring about the performance of locomotion and all other conscious activities.

At a different level the autonomic nervous system operates together with the involuntary muscles it controls. The centre for this control is in the medulla and it involves the co-ordination of such activities as heart beat, respiration, movement of food in the intestine, etc.

The autonomic system may best be thought of as two antagonistic sub-systems. The first of these, called the parasympathetic system, stimulates all activities concerned with normal digestion and "relaxed" running of the body. The opposite system is the sympathetic and this comes into play in emergency situations. Stimulation of this system leads to cessation of digestion, increase in heart beat, respiration and supply of blood to the skeletal muscles. The sympathetic system is activated by the hormone adrenalin.

14. Summary. In this chapter we have seen how both animals and plants respond to stimuli in their environments leading to adaptive changes geared to survival. In animals fast co-ordination is brought about by the nervous system while slower long-term effects are due to chemical co-ordination by the hormones. Plants use only hormonal co-ordination.

A further marked difference is in the effector systems of animals with their complex arrangement of muscles bringing about fast movement of the whole or parts of the body. Plants can only react slowly to stimuli by means of the differential growth movements termed *tropisms*.

It is probably true to say that it is in their co-ordination and response to stimuli that plants and animals differ most compared with any other aspects of their physiology.

PROGRESS TEST 9

1. To what stimuli do organisms respond? (1)
2. What is the role of auxin (hormone) in plant responses? (2)
3. What layers are found in the eye, in order, from the conjunctiva inwards? (3)
4. How does the retina detect light? (3)
5. What happens to sound in the middle ear? (4)
6. How is balance maintained? (4)

7. What can the tongue detect? (6)
8. What parts are found in a nerve cell? (8)
9. What is the sequence of events in a reflex arc? (9)
10. What is the role of the cerebrum of a mammal? (10)
11. What hormone is produced from (a) the thyroid gland, (b) the adrenal bodies? What are the respective functions of these hormones? (11)
12. Why are muscles found in antagonistic pairs? (12)
13. What is the autonomic nervous system? (13)

EXAMINATION QUESTIONS

1. Explain what is meant by the term "response to stimuli." Describe the structure of either the eye or the ear of a mammal. (O. & C., 1969)

2. Show by experiment the action of auxins in plant shoots. (after Cambridge)

3. By means of a large labelled diagram show the main structures of the mammalian spinal cord, including a spinal nerve as seen in transverse section. Indicate the path taken by a nerve impulse from the moment it is generated in the hand when holding a hot plate until the hand is moved. (A.E.B., 1970)

4. The graph shows the effect of the growth hormone on rats. This hormone is produced by the pituitary body at the base of the brain. (N.B. growth curve, Chapter 1)

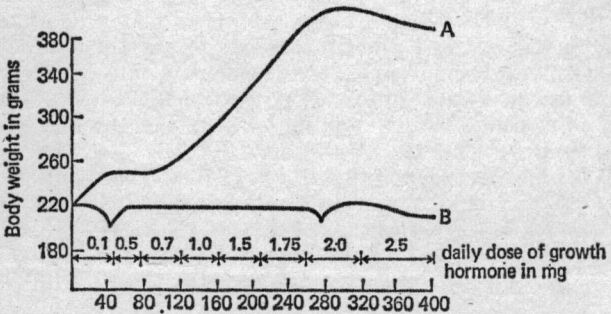

Twenty-four fully grown rats with an average weight of 220 g had their pituitary bodies removed. Twelve were kept as a control (line B) while the remainder were injected with daily doses of the growth hormone (line A). The size of the dose was periodically increased as shown on the dose scale.

(*i*) What in general terms happened to the average weight of the control rats over the period of the experiment?

(*ii*) What was the maximum dose of growth hormone that had an effect in increasing the weight of the injected rats?

(*iii*) During which period was there no increase in weight in the injected rats?

(*iv*) How did the average weight of rats injected during this period compare with their average weight at the start of the experiment? (*A.E.B.*, 1970)

5. (*a*) What would be the effect on the hearing of a mammal if

(*i*) the malleus (hammer) and incus (anvil) are destroyed as a result of infection;

(*ii*) the eardrum becomes heavily thickened as the result of an infection;

(*iii*) the Eustachian tube becomes blocked by mucus?

(*b*) Clearly explain how the inner ear of a mammal provides a sense of balance.

(*c*) A man is sitting in a brightly lit room reading a book. He then goes outside where there is very little light. Why would he be well advised to wait for several seconds before walking down the road? (*A.E.B.*, 1971)

6. Describe the sequence of events in lifting a pencil from the table. (*after Cambridge*)

CHAPTER X

REPRODUCTION OF CELL AND ORGANISM

CELL DIVISION

1. Mitosis. All embryonic cells have the ability to divide and this capacity is carried on into the undifferentiated cells of the adult which are involved in new growth. Cell division which is an exact copying of the parent cell into two daughter cells is called *mitosis* and is common to all types of living organisms: (details shown in Fig. 34).

Man.	Cells dividing by mitosis in mature individual	Malpighian cells of epidermis. Epithelial cells of gut. Bone marrow cells, etc.
	Mature cells no longer able to divide	Muscle cells, nerve cells, red corpuscles, white corpuscles, bone cells, etc.
Plant.	Cells dividing by mitosis in mature state of plant.	Cells of apical meristems. Cambial cells, cork cambium cells.
	Mature cells no longer able to divide.	Xylem and phloem tissue, parenchyma, cork fibres, etc.

Wherever it occurs mitosis is found to have common stages and features and these take place as follows:

(a) *Interphase.* The nucleus of the cell contains chromosomes in pairs, one of each pair having been derived from each parent. During interphase the cell may be active in protein synthesis but the chromosomes are hydrated and do not undergo changes. This is also called the "resting phase."

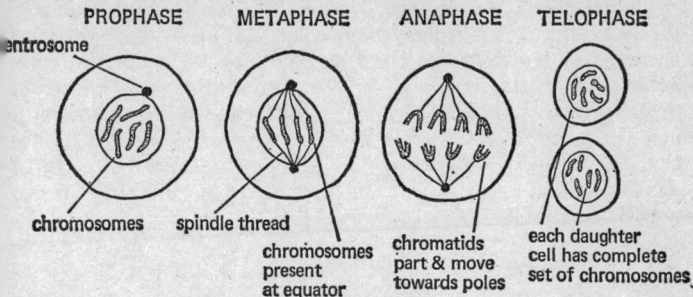

PROPHASE   METAPHASE   ANAPHASE   TELOPHASE

centrosome

chromosomes   spindle thread

chromosomes
present
at equator

chromatids
part & move
towards poles

each daughter
cell has complete
set of chromosomes

Fig. 34. Mitosis.

(b) *Prophase.* During this starting phase of cell division the DNA of the individual chromosomes doubles up and the whole mass of chromosome material gets dehydrated so that it can be readily stained. When an individual chromosome in late prophase is examined it is seen to be a double structure so that we have a cell whose chromosomes are present in pairs and each of which is also paired. During prophase the nuclear membrane is still distinct.

(c) *Metaphase.* As the nuclear membrane begins to disintegrate, the individual chromosomes become attached to spindle fibres in the equatorial regions of the cells. Each chromosome has a centromere which is its region of attachment. The spindle fibres form from the centrosome as shown in the diagram.

(d) *Anaphase.* Here the two DNA strands of each chromosome (which are termed the *chromatids*) part from and repel each other and the two strands move apart, one towards each pole. Anaphase is the most rapid phase of mitosis. There is no nuclear membrane present.

(e) *Telophase.* The two complete sets of half chromosomes (or chromatids) are now pulled towards the poles of the cells by contraction of the spindle fibres. Towards the end of mitosis the nuclear membrane reforms around the chromosomes and the cytoplasm of the whole cell becomes divided into two.

Thus at the end of mitosis a single cell has formed into two identical daughter cells each with the diploid (or double) set of chromosomes.

2. Meiosis. This a special type of cell division that occurs in diploid organisms when they come to make sex cells or gametes or in plants as they make spores. One of its characteristics is the reduction in the chromosome number from diploid (2) to haploid (1), and the other is the interchange of genetic material between two sets of parental chromosomes that takes place. There are eight stages of meiosis but most of the features of interest occur in the first stage. Crossing over of chromosome material is shown in Fig. 35.

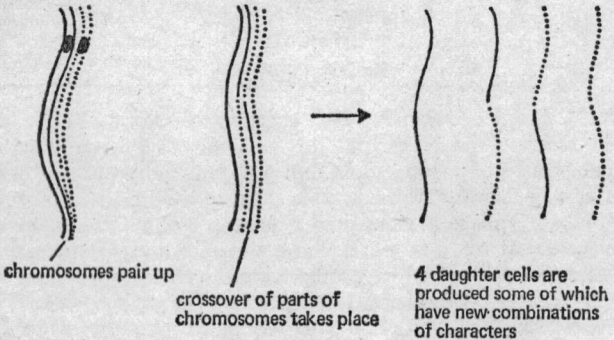

chromosomes pair up

crossover of parts of chromosomes takes place

4 daughter cells are produced some of which have new combinations of characters

FIG. 35. Meiosis, showing chromosome crosses.

(a) *Prophase 1.* Moving from interphase, the non-dividing part of the cell's life, sex and spore mother cells enter on the first active stage of meiosis, that is, prophase 1. The chromosomes appear as long thin strands which at first are single. These come together in pairs and each strand thickens and doubles up and each forms into two identical chromatids.

Towards the end of the phase the four strands of each pair now become very closely intertwined and pieces of chromatid are exchanged between the pairs. This is very important as it is the basis of genetic variation between parent and offspring.

(b) *Metaphase 1* is rather similar to the metaphase of mitosis but in this case the chromosomes are working together as complete pairs which move to the equator of the cell to take up positions on the newly constituted spindle fibres. Once again the chromosomes are attached to these by the centromeres. It should be noted that superficial examination of the cell appears to show only half the number of chromosomes present as each apparent chromosome is made up of a parental pair.

(c) *Anaphase 1*. Individual chromosomes of each pair now move away from each other but during prophase some pieces of material have been exchanged so that these are not the original chromosomes but new arrangements of DNA.

(d) *Telophase 1* sees the division of the cell into two each of the daughters, now having half the number of chromosomes of the original mother cell, as well as having newly arranged chromosomes. In fact the really important events of meiosis have occurred by the end of this phase.

Sometimes there is a brief resting period between telophase 1 and prophase 2 and sometimes they move straight from the former to the latter.

(e) *Prophase 2*; now the chromosomes reappear, each being formed of two chromatids joined at a centromere.

(f) *Metaphase 2*. The equatorial plane of the cell has now changed through 90° and in each of the daughter cells, chromosomes take up their metaphase positions at the respective equators of the cell. The chromosomes have a centromere which again becomes attached to the newly formed spindles.

(g) *Anaphase 2*. All the chromosomes involved split into chromatids and these move towards the poles along their fibres (*see* Fig. 34)

(h) *Telophase 2* is the last phase of meiosis and here all the four new daughter cells of the original parent cell have their nuclear membranes reconstituted. It will be noted that each of the four daughter cells has only one set of chromosomes, as compared with the two sets originally present. As important as this is the fact that the one set that it does possess are in fact mixtures of the original parental genetic material.

(i) *Significance of meiosis* lies in the halving of the chromosome set so that two gametes each carrying half a set of chromosomes will, on fusing to make a zygote, reconstitute a complete set in the offspring.

The other important point is in the variation or "shuffling" of genetic material that is a feature of the process. It means that no gamete is carrying exactly the same combinations of genes as its parents and therefore variation will result in the progeny.

This variation allows the process of natural selection to operate and the production of variable offspring may be regarded as the main biological purpose of sexual reproduction.

SEXUAL REPRODUCTION

3. Definition of sexual reproduction. Sexual reproduction is a type of production of progeny that involves the fusion of gametes or sex cells. In higher organisms these are usually different and in animals consist of a minute motile sperm and a larger food-laden egg which remains sessile. In higher plants the sperm has been reduced to a male nucleus which is transported to the vicinity of the egg in a growing pollen tube. The egg is little more than a single nucleus within the ovary of the flower.

It should be noted that all higher animals are either male or female so that self-fertilisation cannot occur. The same is not true for the great majority of higher plants which are in fact hermaphrodite, having both male and female organs. In plants therefore it is possible for self-fertilisation to occur although this is normally discouraged by one mechanism or another.

As will be seen in **16** it is biologically advantageous to have outbreeding rather than inbreeding, as the former not only produces more variation but lowers the tendency of combining harmful characters or genes in a double dose.

SEXUAL REPRODUCTION IN THE PLANT

4. The flower. The organ of sexual reproduction in the green plant is the flower. This typically consists of sepals and petals which surround the male and female parts. All the parts of a flower have some functional significance and these can be considered in sequence for a simple flower such as the buttercup (*see* Fig. 36).

(a) *The sepals* are leaf-like and enclose the flower bud, providing protection from desiccation, etc., when the flower is immature. They are comparatively much smaller than the

The parts of a flower such as a Buttercup

petal

anther
filament } stamen

carpel

receptacle

sepal

The process of pollination consists of the transfer of pollen

pollen grains germinate on stigma

pollen tube grows along the style and ovary wall

ovule, surrounded by soft integuments

female nucleus

male nucleus released from the end of the pollen tube fuses with female nucleus to form a zygote. This is fertilisation

The formation of the fruit

the ovary wall becomes leathery — it makes up the outside of the fruit

the integument becomes the protective seed coat (testa)

within the seed is the embryo new plant and its food reserve

the whole ovule is now the seed

the fruits will break off from the receptacle and can be blown short distances which assists the dispersal of the plant.

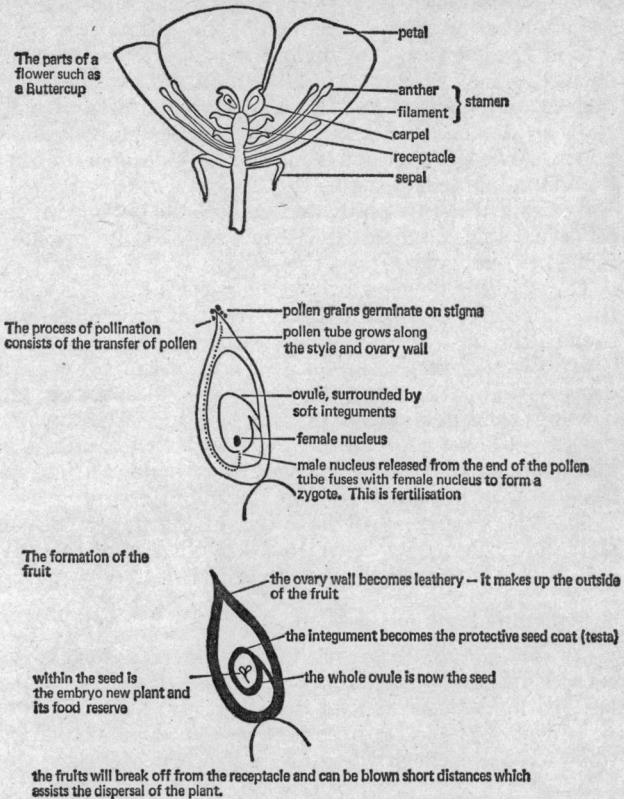

FIG. 36. Reproduction in the buttercup.

petals and are green in colour. In the buttercup there are five.

(b) *The petals* are more specialised and in the mature flower they are large and yellow in colour as well as having a definite bright sheen. This combination of factors makes them very conspicuous to bees and other pollinating insects which are attracted to visit the flower. At the base of each petal is a nectary which secretes sugar and this is sucrose, which makes up part of the food of the pollinator.

(c) *The stamens* are the male part of the flower, and in the buttercup they are found in large numbers surrounding the carpels or female parts. Each stamen consists of a long stalk or filament and a head or anther which contains the pollen. When the anther is mature it splits open to release the individual pollen grains.

In flowers that are pollinated by bees the pollen grains are relatively large and sticky so that they easily become attached to the hairy legs of the insect.

The stamens develop before the carpels and this reduces the chances of self-pollination, although in simple flowers such as the buttercup this must often occur.

(d) *The carpels* are the female parts of the flower. They consist of an oval structure, the carpel wall, inside which develops the single female nucleus or ovule. The top end of the carpel has a short neck or style which terminates in a tacky stigma adapted for the receipt of pollen. There are a large number of carpels present in the buttercup.

(e) *The receptacle.* All the parts of the flower described above are based on a central core at the top of the flower stalk. This is termed the *receptacle.*

5. Pollination and fertilisation.

(a) *Pollination.* The whole buttercup flower is very conspicuous and adapted to attract insects such as bees. A bee visiting a buttercup for its nectar and pollen will get a good deal of the latter over its legs and abdomen. Besides this the bee specifically brushes pollen into sacs on its hind legs.

It is possible that a particular buttercup visited will have mature stamens in the outer whorls while the inner ones are still immature and curl over the carpels. A bee collecting pollen from such a flower may then visit an older flower whose stigmas are receptive and in crawling over this flower will transfer to it pollen from its previous collecting.

If this pollen is deposited on the stigma it will germinate. The transfer of pollen from one flower to another of the same species in this way is called *cross pollination.*

(b) *Fertilisation.* The pollen grains on the stigma germinate to produce a long tube which grows down through the style and carpel wall until it reaches the entrance of the ovule. Here a male nucleus is released and this enters the ovule via the micropyle and fuses with the egg nucleus within. The fusion of these two nuclei is called *fertilisation*.

6. Seed and fruit formation. After fertilisation a number of changes take place in the flower which lead to the formation of seeds and fruits while the sepals, petals and stamens, which have all served their purpose, now wither away and drop off the expanding carpels. A variety of fruits are shown in Fig. 37.

(*a*) *Seed formation.* The fertilised egg nucleus is now termed a *zygote* and it begins to develop into a tiny embryo plant with a minute root and shoot and two first leaves called *cotyledons*. At the same time that this is happening the layers around the ovule become hardened into a testa or seed coat which protect the developing embryo. Within the ovule food reserves have been accumulated from the parent plant and these will later be mobilised on germination.

The embryo with its food reserves and protective coat is now termed the *seed* and this will eventually be dispersed away from the parent plant and germinate and grow into a new buttercup. The means of its dispersal is the fruit.

(*b*) *Fruit formation.* At the same time as the seed is developing changes are taking place in the carpel wall. In the case of the buttercup it becomes enlarged and then dries off until it will eventually fall or be blown off the receptacle. The chances are that it will be carried at least a short distance from the original plant and thus prevent direct competition between the developing seed and its parent for space and light. Buttercups do not have very specialised fruits and their dispersal is extremely limited. Many more advanced fruits which develop wings for wind dispersal or fleshy walls to attract animals to eat them, or are sticky, or have hooks to cling to animal fur are capable of dispersing seeds great distances from the original parent plant. Even the familiar "explosive" pod of the pea family (*e.g.* gorse) is capable of throwing seeds several feet away from the parent.

7. Germination of the seed. Plants produce far more seeds than they would need to maintain their numbers if all the seeds germinated. In fact only a proportion (rather less than

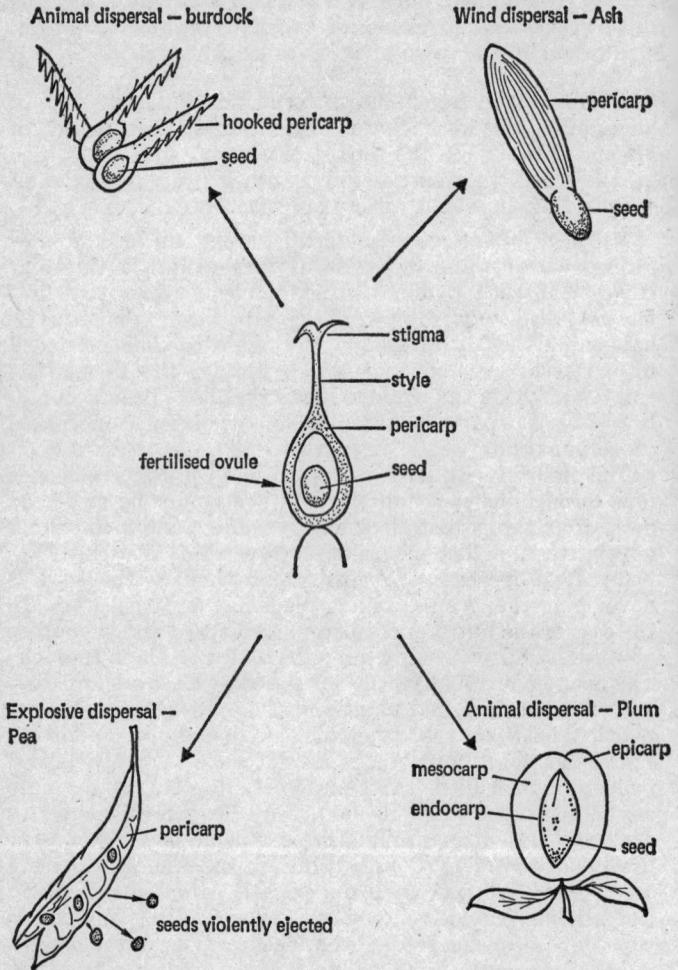

Animal dispersal — burdock

hooked pericarp
seed

Wind dispersal — Ash

pericarp

seed

stigma
style
pericarp
fertilised ovule
seed

Explosive dispersal —
Pea

pericarp

seeds violently ejected

Animal dispersal — Plum

epicarp
mesocarp
endocarp
seed

Fig. 37. The formation and dispersal of seeds and fruits.

one per cent) tend to survive for one reason or another. Most seeds have a built-in "clock" and will not germinate directly on dispersal but must go through a period of dormancy of varying length. Once this period is passed the seed will germinate provided it has warmth, moisture and oxygen present. Whether or not it will survive after germination and the utilisation of its food resources depends on its surroundings and ability to withstand competition from other species in its immediate vicinity. The stages in germination in a seed (such as the broad bean) are as follows:

(a) *Water uptake and changes in metabolism.* The dry seed has a high osmotic potential and placed in moist conditions it will start to draw in water through the permeable micropyle by osmosis. This in turn leads to the hydration of the cells and the hydrolysis of the stored food, which is normally starch with some protein.

The rate of respiration greatly increases and the apical regions of the young root (radicle) and young shoot (plumule) begin to grow.

(b) *Emergence of the radicle.* This happens next as the food reserves are translocated into the radicle which rapidly expands and forces its way out through the testa. The radicle is positively geotropic (*see* page 105) and grows down into the soil. Within hours it begins to put out root hairs and these increase the rate of water uptake by the germinating plant. Later lateral roots are put out.

(c) *Emergence of the plumule.* A day or two after the radicle has emerged the plumule or shoot starts its expansion and forces its way out through the split testa. Food substances are now being translocated from the reserves in the cotyledons to the apical meristem of the shoot. Unlike the radicle the plumule is negatively geo- and positively photo-tropic so that it grows upwards through the soil into the light and air.

(d) *The new plant established.* A germinating seedling has only a limited food reserve and after a couple of weeks the first leaves above the ground will have turned green and started to photosynthesise.

At this point the young plant becomes self-supporting as newly synthesised food materials pass down into the shoot and root system.

REPRODUCTION IN THE MAMMAL

8. The male. The reproductive processes of all mammals have much in common, but for our purposes that of the human will be taken as it has certain special features which seem particularly important in the biological success of our own species. We may start with a consideration of the anatomical features of the two sexes, that is their primary sexual characteristics.

(a) *The testes.* These are the organs where the spermatoza and male sex hormones are produced. Each testis consists of many feet of coiled tubes from the walls of which

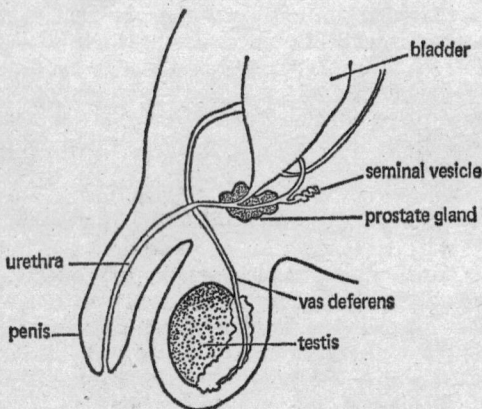

FIG. 38. The human male reproductive system.

sperm cells are budded off by the process of meiosis. In between the tubes are the glandular cells that manufacture testosterone.

Sperm collects up in the complex tubular arrangement of the vas efferens at the base of the testis and thence travels via the large single vas deferens to the top of the urethra. Glands opening into this include the seminal vesicles and the prostates, and these provide nutritive fluid and additional hormones.

The sperm shares a common duct down the urethra which is the tube running down the penis. The latter is a spongy organ which when filled with blood can be inserted into the vagina of the female, thus allowing internal fertilisation to take place. The anatomy of the human male sex organs is shown in Fig. 35.

(b) *The sperm cells* are produced in millions and indeed the chances of any particular sperm arriving at the egg are so small that it is necessary for very large numbers to be present for any hope of fertilisation. The front end of the individual sperm is a nucleus which carries a single (rather than a double) set of chromosomes. Behind this is the middle piece which acts as a power pack to provide energy for the whipping motion of the tail which propels the sperm along. The sperm has a positive chemotaxis towards the egg and in terms of its own length (it is only 0·01 mm) a considerable distance to swim to reach its object.

9. The female has a more elaborate sexual anatomy than the male and a complex interaction of hormones that ensure regular cyclic production of the eggs from alternate ovaries. She is also more deeply involved in parental care of the off-spring than the former, although the roles of the two sexes may be regarded as complementary in reproduction.

(a) *The ovaries* are small organs a few centimetres in length that lie at the top of the oviducts, one ovary to each side of the body. Each ovary has an enormous number of potential eggs, somewhere in the region of 30,000, but only one egg develops at a time and the majority never come to maturity at all.

The developing egg lies in a Graafian follicle which gradually expands under the influence of the follicle-stimulating hormone from the pituitary. After fourteen days the egg is mature and is released from the follicle into the fallopian tubes at the top of the oviduct.

The mature egg is approximately 0·2 mm across and it contains besides the nucleus a small reserve of yolk.

(b) *The uterus* is a muscular organ into which the two oviducts lead. It has an epithelial lining which goes through a cycle of growth and shedding each month. The uterus is capable of great expansion to some 500 times its normal size and its muscles are extremely powerful, as they need to be in the final expulsion of the foetus at birth.

(c) *The vagina* is an ante-chamber to the uterus and it i.
here that the sperms are deposited during copulation. A
the mouth of the vagina is the clitoris, a sensitive organ and
the homologue of the penis. The anatomy of the human
female sex organs is shown in Fig. 39.

(d) *The mammary glands.* Besides the primary sex organs
described the female has a pair of mammary glands whose
cyclic activity is linked to that of the ovary. If fertilisation
occurs and a foetus starts to grow inside the uterus the

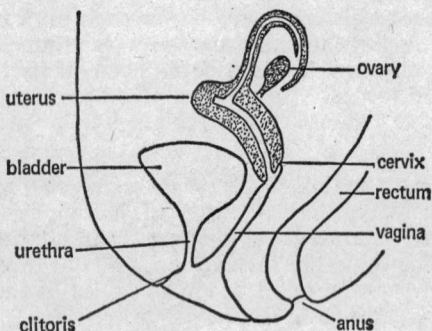

FIG. 39. The female reproductive system.

mammary glands become enlarged and near the time of
birth begin to secrete milk, which will actually be released
shortly after the baby is born.

10. Fertilisation and development of the foetus. If copulation takes place at that time in the female menstrual cycle
when the egg is in the oviduct then there is a probability of it
coming into contact with many sperm. These are attracted to,
and surround the egg but only a single sperm can actually
enter the egg membrane which thereafter becomes quite impermeable to further sperm penetration. The fusion of the
sperm and egg is called *fertilisation* and the product is termed
the *zygote*.

(a) *Fate of the zygote.* Within a few hours of fertilisation
the zygote has formed many cells and is now termed an
embryo or *foetus*. At the same time it develops tiny villi

along its outer margins and these serve to anchor it to the side of the uterus. For a short period nutritive substances pass from the uterus lining into the embryo via the villi.

(b) *Development of the placenta*. A more permanent and much larger attachment between the foetus and the uterus is the placenta and this develops from both maternal and foetal tissues. The placenta has a very large surface area in close contact with the mother's bloodstream while on the foetal side it leads via the umbilical cord to the navel.

Although the two bloodstreams are not actually in contact it is possible for small molecules to pass across the placenta. From the mother's blood pass oxygen, water and all the ingredients of diet necessary for the growth of the foetus.

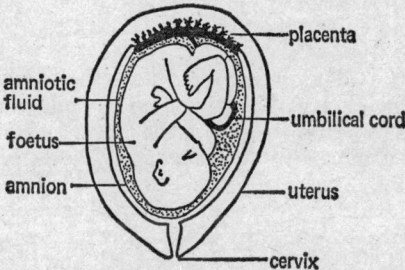

FIG. 40. The foetus inside the uterus.

From the foetal blood pass out waste carbon dioxide and nitrogenous excretory matter such as ammonia and urea.

(c) *The amnion*. While the foetus is beginning to grow and differentiate it develops around itself a thick membrane or amnion. Within this the amniotic fluid is secreted and in this the foetus floats. The amnion and its fluid protect the developing foetus from desiccation, mechanical damage and infection. The relationship of these organs is seen in Fig. 40.

(d) *Development of the foetus*. The original mass of cells produced immediately after fertilisation soon begins to differentiate and an elongated three-layer embryo is produced. At first this has a large head, gill slits, no limbs and a long tail and is rather similar to a fish. Gradually however these features are lost and by the end of the third month of pregnancy the foetus resembles a human baby, although at this stage it is still only some 6 cm in length.

During the remaining period of gestation, and especially during the last three months, the foetus grows rapidly until just before birth it may be some seven pounds in weight. During the whole of this period it has been nourished and supplied with oxygen via the placenta which has now come to be about the size of a soup plate.

11. The process of birth. Sometime during its last few weeks inside the uterus the foetus turns upside down so that its head is near the cervix or opening of the uterus. The actual process of birth is instigated by a pituitary hormone which causes the uterus to contract rhythmically. The contractions move down from the top to the bottom of the uterus and tend to push the amnion hard against the cervix and underlying pelvic girdle.

Eventually the amnion ruptures and the release of the amniotic fluid together with further contractions of the uterus cause the foetus to be expelled from the mother's body. It will however still be attached to the placenta by the umbilical cord.

Within minutes of birth the baby has expanded its lungs and begun to breathe air and the umbilical cord is severed by the mother (or doctor in our own case). The placenta becomes detached from the uterus shortly after birth and is expelled as the "afterbirth."

12. Parental care. The newly born baby is very helpless, especially in the case of humans, and for a period of months it will feed from its mother's milk and be completely dependent on her for its survival. We call this phenomenon *parental care* and while it is present in all mammals it is especially developed in our own species where it may be described as lasting for a number of years.

During this critical period of development there is a close bond of communication between parent and offspring and a high capacity on the part of the latter to learn. Because of the efficiency of the parental care the survival chances of the young human are very high and this is perhaps one of the reasons why we produce only one offspring at a time.

VEGETATIVE REPRODUCTION

13. Definition of vegetative reproduction. Unlike sexual reproduction vegetative increase does not involve the fusion of gametes. In this process a new individual is formed from a part of the original parent plant (the process does not occur at all in higher animals). Eventually the new plant separates from the original parent.

The main advantage of vegetative increase is that it is a very safe form of reproduction. The developing individual can be maintained by food from the established parent and it does not depend on the chances of insect visits or successful dispersal of fruits. The disadvantages of the process are twofold, firstly that the dispersal of the offspring is very limited, seldom more than a few feet from the parent so some competition between the two must occur; secondly there is no genetic interchange so that vegetatively produced plants exactly resemble the parents in all their characters. This in the long run would be a very serious disadvantage as it would decrease the ability of the species to adapt to new conditions.

In fact a very large number of plants combine both processes of sexual and vegetative reproduction in their life histories, thus getting the short-term advantages of the latter and yet retaining the genetic potentialities conferred by the former.

Vegetative propagation is often combined with perennation or survival over winter as well as with some degree of dispersion. From many possible examples the following illustrate the nature of the process.

14. Examples of vegetative reproduction.

(a) *Strawberry runners.* A mature strawberry plant sends out a number of shoots that grow along the ground horizontally. Where these form nodes they put down adventitious roots while an aerial shoot forms from the node region. After the young plant is established its connection with the parental stock withers away so that a self-supporting new individual is formed. A single mature plant may produce a dozen or so runners in this way.

(b) *Rhizomes* are underground stems which grow horizontally. They send up new shoots from lateral or terminal buds and produce adventitious roots at the nodes. Food reserves are laid down in the stem region so this is a means of perennation as well as vegetative propagation. Common examples are irises and nettles but the method is also common in herbaceous plants.

(c) *Tubers* such as potatoes are a very successful modification of the rhizome. The end of the underground stem becomes greatly swollen with food and despite its position and its bark covering, it can clearly be seen to be a stem structure as it has scale leaves and terminal and lateral buds.

The tuber is a means of multiplication as well as of the dispersal and perennation of the original plant. It should be noted that all the plants mentioned above as examples of vegetative propagation have effective sexual methods of reproduction as well whereby seeds and fruits are produced.

A NOTE ON GENETICS AND EVOLUTION

15. How characteristics are inherited. As mentioned in Chapter II, the DNA of the nucleus carries the genetic or heredity instructions in the form of genes. These influence and control the formation of proteins within the cell, and each individual organism will have its own peculiar set of genes.

In sexual reproduction similar or contrasting genes come together from different individuals. If the genes are the same then the offspring will show (for the particular character controlled by the gene) exact similarity with the parent. If however the genes inherited by an individual have different effects then one will be dominant, *i.e.* will show its effect, while the other will be recessive and its effects will not be seen. An individual with similar genes is said to be *homozygous* while one with dissimilar ones is said to be *heterozygous*.

In mice the colour grey or agouti is the normal type and is dominant over the albino white. If we have pure lines of mice of the two colours and these are crossed we would get the following result

Parents: Grey X White
Offspring: All Grey

If these offspring are then mated among themselves it will be found that the progeny are grey to white in the ratio of 3 : 1. It is customary to give the dominant gene the capital letter of the character it controls and the recessive gene the small letter so that (bearing in mind that the chromosome number is halved at gamete formation) we can set out the above crosses in symbols as follows:

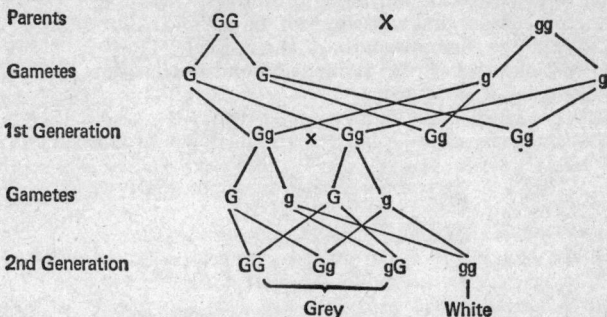

Parents	GG	X	gg
Gametes	G G	g	g
1st Generation	Gg x Gg Gg Gg
Gametes	G g G g
2nd Generation	GG Gg gG gg
	Grey	White

This basic understanding of the nature of inheritance was first worked out on garden pea plants by Gregor Mendel in 1870. It shows us why breeding individuals from different stocks produces variations. In the 1930s it was found that genes could change by a process of mutation to produce quite different effects, such mutations being caused by radiation or by chemicals or even happening spontaneously.

16. Genetics and evolution. By far the majority of mutations are harmful but every so often one will turn up that confers some advantage on its owner. For this reason he or she will tend to be more successful and have more progeny than other members of the species. The new advantageous gene thus rapidly passes into the population by means of sexual reproduction and so in a number of generations it may come to be possessed by most of the population. In this way and over very long periods of time species change and become more highly adapted to their environments.

We call this process of gradual change *evolution* and it is a continuous although extremely slow process from which even man is not exempt. Clearly it is important for the survival of a given species that it should evolve and equally clearly the importance of sexual reproduction in the spreading of favourable genes can be seen. It is not surprising that sexual processes are highly developed in both the mammals and green plants.

PROGRESS TEST 10

1. Where is mitosis likely to be found? **(1)**
2. How does meiosis differ from mitosis? **(2)**
3. How may sexual reproduction be defined? **(3)**
4. What are the male parts of the flower? **(4)**
5. Distinguish between pollination and fertilisation. **(5)**
6. What is a seed? **(6)**
7. What is the function of a fruit? **(6)**
8. Why does a seed planted upside down produce a normal plant? **(7)**
9. Where are the sperms formed in a mammal? **(8)**
10. What is the Graafian follicle? **(9)**
11. What is the function of the placenta? **(10)**
12. To what extent is parental care developed in mammals? **(12)**
13. How may a rhizome be defined? **(14)**
14. What is a gene? **(15)**
15. What is a dominant gene? **(15)**
16. What did Mendel show? **(15)**

EXAMINATION QUESTIONS

1. What is pollination? What is the significance of pollination in the reproductive processes of a named plant? Describe the principal ways in which pollination may take place. (*O. & C.*, 1970)

2. Briefly describe parental care in a named mammal. How does the young obtain food (*a*) before birth, (*b*) shortly after birth, (*c*) at maturity? (*O. & C.*, 1970)

3. Define pollination and fertilisation. Describe one of these processes in a named plant. (*A.E.B.*, 1970)

4. Describe how a cell divides in the apex of a stem. (*after A.E.B.*)

5. Describe the germination of a named seed. What are the conditions necessary for germination to occur? (*O. & C.*)

6. Use one word only in each box to label the structure indicated, the missing word or the process referred to.

In cell division the process of produces two identical cells.

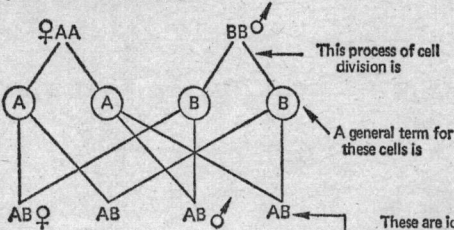

This process of cell division is

A general term for these cells is

These are identical in appearance to the female parent *AA*.

Extend this diagram in this space to show how the next generation is produced and the range of offspring possible.

The Factor *A* is said to be

The Factor *B* is said to be

In plants *A* would be in the

B would be in the

(after O. & C.)

CHAPTER XI

REPRESENTATIVE LOWER ORGANISMS

SINGLE-CELLED ORGANISMS

1. Algae. Although there are many-celled algae such as the seaweeds, the majority of the group that are important in the life of higher organisms are either single-celled or colonial forms.

(*a*) *Characteristics of algae.* The algae are the simplest forms of photosynthetic plant. They are mainly found in water, where they form the primary producers absorbing CO_2 and salts and manufacturing organic substances.

Algae provide food for herbivorous members of the zoo-plankton such as many crustacea, and they are to aquatic food chains what grass and trees are to terrestrial ones.

At the present time algae are not directly important to man as a food source although some attempts have been made to crop and process them directly. The simpler forms are however of some biological interest because of their very low grade of organisation and because they may share both plant and animal characteristics.

(*b*) *Evolutionary significance of algae.* It seems probable that the simpler green algae are very close to the original life forms on earth and that animal and fungal forms as well as bacteria evolved from this group a very long time ago. Because of the very large numbers of algae in the surface waters of lakes and the sea, and the photosynthesis they undergo, this group is actually more significant in the return of oxygen to the atmosphere than the photosynthesis of land plants.

A very simple green alga is the freshwater Chlamydomonas (*see* Fig. 41). Because it carries out photosynthesis and stores starch and has a cellulose wall it is taken to be a plant. Unlike higher plants however it is capable of active swimming.

142

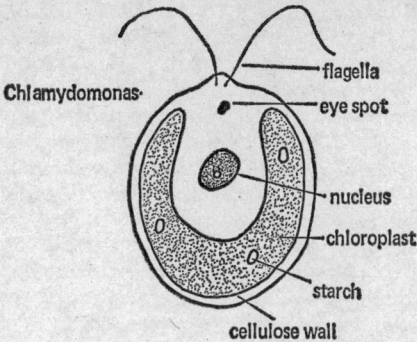

FIG. 41. Unicellular green algae.

Algae such as Chlamydomonas reproduce sexually by the fusion of similar gametes and asexually by fission into a number of smaller individuals.

2. Protozoa are definitely the most simple members of the animal kingdom as they cannot photosynthesise and depend on plants for their food. They do not have cellulose walls or store starch. They are all unicellular organisms (*see* Fig. 42). The

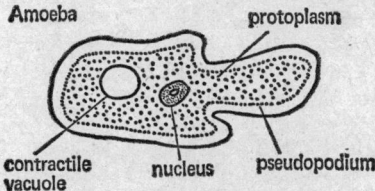

FIG. 42. A protozoan (amoeba).

importance of protozoa to higher organisms is various, both beneficial and the reverse.

(*a*) *Beneficial aspects of protozoa.* Many protozoa feed on algae and are the next stage up in food chains that eventually lead to higher organisms. Protozoa of many species are present in great numbers in soil where they assist in the breakdown and return of organic substances.

The ruminant mammals, that is, those that feed on vegeta tion which they regurgitate and chew again, such as shee and cows, have large numbers of protozoa (and bacteria) i the intestine, and specifically in the rumen. These minut organisms are vital to the digestive processes of their host as they have the enzymes for the hydrolysis of "difficult plant substances such as cellulose as well as the ability t synthesise vital nutritional substances. This relationship c mutual benefit is termed *symbiosis*.

(*b*) *Deleterious effects of protozoa.* Certain protozoas als live within the tissues of other animals but in this case ar there as parasites and cause a variety of diseases. In ma the most common parasitic protozoa are malaria, dysenter and sleeping sickness, although there are a number of others

These parasitic protozoa are often carried by means o vector insects such as mosquitoes, or various sorts of flies, and elimination of the vector does a great deal to control th disease itself.

Parasitic protozoa have very considerable powers of asexual reproduction (typical of parasites in general) which helps make up for the hazardous nature of their existence.

3. Bacteria are unicellular or colonial forms and may be a remote offshoot of the fungi, with whom they share certain characteristics (*see* Fig. 43). For the most part they are dependent on plants, being unable to photosynthesise. The bacteria have very high rates of reproduction and form extremely resistant spores which are present in the soil and atmosphere of the earth (*see* Chapters III and VIII). They are also found in the sea. They range from 1–25 μ ($\mu = \frac{1}{1000}$ mm) in size and are usually round (cocci), rod-shaped (bacilli) or spiral (spirelli) in form.

(*a*) *Advantageous bacteria.* By far the majority of bacteria are saprophytic and feed from dead organic matter which they digest outside their bodies and assimilate in liquid form. Bacteria are the main saprophytes in the essential nutrient cycles which break down organic matter in the soil and sea bed and return to circulation. Like the protozoa described above many bacteria live as symbiotes in the intestines of higher animals such as ruminants.

Bacteria have great numbers of different enzymes which

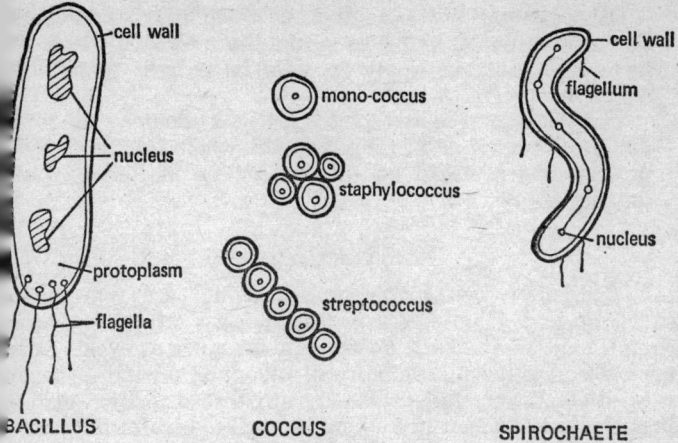

cell wall

nucleus

protoplasm

flagella

BACILLUS

mono-coccus

staphylococcus

streptococcus

COCCUS

cell wall

flagellum

nucleus

SPIROCHAETE

FIG. 43. Types of bacteria.

they can produce, and except for modern man-made fabrics, they will eventually hydrolyse such stable substances as leather, wool and bone, as well as cellulose and lignin and cork.

The fermentation properties of certain bacteria have been used in the manufacture of cheese and other foods. Symbiotic root nodule bacteria have been described in connection with the fixation of atmospheric nitrogen.

(b) *Harmful bacteria.* There are a considerable number of parasitic bacteria that cause disease in man and all mammals suffer from specific bacterial infections. Examples of human diseases caused by bacteria are forms of dysentery and pneumonia, sore throats, boils, typhoid, tetanus and tuberculosis, to mention only a few.

In recent years the way in which such bacteria are spread has led to a great reduction in the number of deaths due to these diseases, and in the decades since the war antibiotics prepared from soil fungi have proved most effective in treating infections. The principles of immunisation by introduction of weakened form of pathogenic bacteria has virtually eliminated such old-time killers as typhoid, cholera and tuberculosis from large numbers of the human population.

Other harmful bacteria are those which infect and kill ou domestic animals, and here again the measures which w have been able to apply to man have been reasonabl} effective.

The spoilage of stored food is brought about by the join action of bacteria and fungi and the whole science of foo preservation is based on eliminating or inhibiting thes(organisms.

THE FUNGI

4. Characteristics of the group. Fungi are non-photo synthetic and mainly a saprophytic class whose activitie complement those of the bacteria in the nutrient cycles. The typical fungus has a cellulose wall and stores oil and glycoger while its body is made up of many tiny threads called hyphae, bound into a stronger mycelium. Fungi are present in the soi]

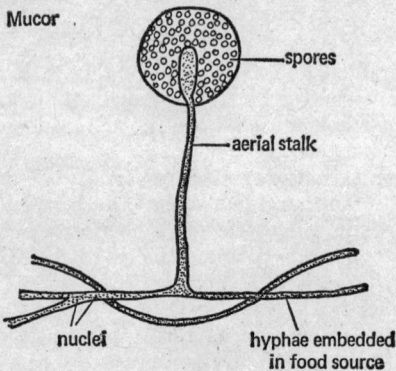

Fig. 44. A fungus mould.

as moulds, and like bacteria they have the ability to break down a whole range of organic substrates.

A typical soil fungus (Chapter VIII also) is the mould Mucor (see Fig. 44). The hyphae of all fungi are permeable to water and they cannot survive in dry conditions. When the nutrient supply in their particular area has been exhausted they will form fruiting bodies which produce hundreds of thousands of resistant spores.

Fungi are great spoilers of stored foods (*e.g.* bread mould) but their enzymic processes are also used beneficially, as in fermentation by the yeasts to produce alcohol. More than bacteria, fungi attack plants as parasites and in the form of blights and rusts and damping off disease cause great losses to crops each year.

In 1931 it was first observed that some soil fungi produce chemicals which inhibit the growth of bacteria and from these early observations our modern range of antibiotics have been developed, *i.e.* penicillin, tetramycin, aureomycin, etc. These chemicals are relatively harmless to mammals but very toxic to the majority of pathogenic bacteria. The larger fungi such as mushrooms can be eaten and contain a rich source of vitamins. Other fungi, as is well known, are extremely poisonous.

THE WORMS

5. Characteristics of the worms. These comprise a number of separate and rather distantly related groups. The segmented worms such as the earthworm stand somewhere between the protozoa and the vertebrates in their complexity and are of evolutionary interest in this respect. The earthworm is a very familiar example and is an extremely beneficial animal because of its effects on soil fertility. Earthworms have besides

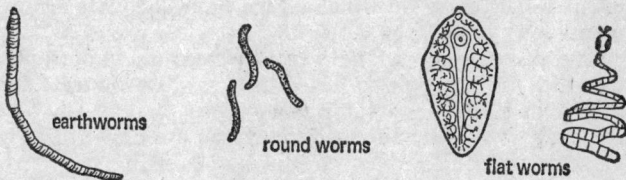

earthworms   round worms   flat worms

FIG. 45. Types of worm.

three layers of cells, a body cavity, and are also segmented; it is almost certain that they gave rise to both the arthropods (jointed limbed animals) and the molluscs (snails, mussels, etc.). Only one group of segmented worms is parasitic—the leeches— and these are blood-suckers that may cause serious infections by their bites (*see* Fig. 45).

(*a*) *The roundworms*, properly called *nematodes*, are

extremely widespread and important as parasites of both plants and animals. The former include such pests as the eelworms which live in the soil and can infest the roots of many plants, causing poor growth or even death. In Britain the round-worm Heterodera is an important pest of potato crops and causes losses of thousands of pounds worth of potatoes each year.

Most animals harbour roundworms as parasites; they are for example almost always present in puppies. Ascaris is a major example of the group and this is an intestinal parasite of the pig which very occasionally gets transmitted to man.

Specifically human parasitic roundworms include Elephantiasis (carried by mosquitoes) and hookworms. Both these parasites can lead to death but on the whole they just produce disability for the infected person. Diseases caused by roundworms are extremely common in tropical regions.

(b) *The flatworms*, so called because of the dorso-ventral flattening of their bodies, have two important parasitic subgroups. The first of these comprises the flukes which tend to parasitise the liver and cause continuous blood loss and weakening. Both the flukes of man (such as Bilharzia) and those of sheep (Fasciola) use a water snail as intermediate host. Through a knowledge of their life cycles it is possible to control them in countries with hygienic sewage disposal. Again liver flukes are commonly found in the poorer countries of the world which do not have adequate arrangements for hygiene.

The other group of flatworms which are important as parasites are the tapeworms. These usually complete their life cycle through an intermediate host and for man the pork tapeworm and the beef tapeworm are likely to be the most troublesome. Where faeces is disposed in such a way that worm cysts can get into livestock and where meat is likely to be undercooked or not inspected, there the tapeworms flourish. It was recently estimated that over 70 per cent of the population of Ethiopia were infected by such worms. In Britain their occurrence would be fractional.

THE INSECTS

6. Characteristics of insects. Insects are the most successful of all types of animals and comprise more than 70 per cent of

he whole number of species that exist. They are characterised
by a tough external skeleton of chitin, by three pairs of jointed
imbs, by a body divided into a head, thorax and abdomen,
and by two pairs of wings. Insects also breathe by direct
diffusion into a tracheal system (*see* Fig. 46).

(*a*) *Beneficial insects.* Most species of insect that exist
do not affect man in any way. On the other hand there
are several species which are extremely beneficial to man
and even more which are extremely harmful in one way or
another.

The beneficial insects are mainly the pollinators such as
the bees and butterflies. These are the agents who produce
cross pollination and thus fruit formation in the majority of
plants that produce coloured flowers. There is really no
human substitute for these pollinators and it is a very
difficult problem that insecticides that are aimed at pest
control also decimate populations of bees.

Other groups of insects that can be regarded as beneficial
are those which parasitise other pest species. Thus ladybirds

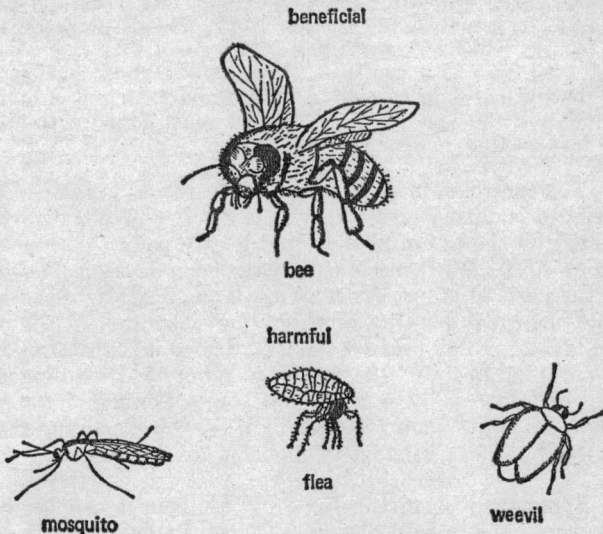

FIG. 46. Types of insect.

are an important natural controller of aphids and ichneumon and other flies keep down the population of cabbage whites.

(b) *Harmful insects.* Insects that are harmful are those that live directly on man as parasites, especially where they act as vectors of disease. There are also the many species that devour our food crops or our food stores. It is not difficult to select examples of each type:

(i) *Parasitic insects.* Some species of mosquitoes suck the blood of man and are responsible for the transmission of diseases such as malaria and yellow fever. Fleas may live on us and cause only irritation by their bites, but in some cases however they can transmit bubonic plague, the "black death" of the Middle Ages. Flies transmit diseases such as food poisoning and dysentery by feeding on garbage and sewage and then infecting our food with harmful bacteria. Blood-sucking flies such as the tsetse of West Africa carry the disease sleeping sickness. Again body lice, little more than irritating by themselves, are the transmitters of typhus.

(ii) Almost every type of plant that humans grow to eat has specific insect pests and it has been estimated that these pests eat approximately one-third of all the food man grows. Common examples are greenfly, which attack beans and many other crops, Colorado beetles, which live on potatoes, and locusts, which eat everything that is green.

(iii) There are also the insects of stored foods such as the weevils and flour beetles. All these take their toll of human crops and some (such as aphids) may actually transmit diseases from one plant to another.

7. The control of insects. In recent years and in most of the developed countries insect vectors of disease have been exterminated by the use of modern and highly potent insecticides such as DDT. Other insecticides have been developed to control the pests of crops, again with a large measure of success. While the great benefits of these new chemicals in the war against famine and disease cannot be denied it is also true that certain side effects of the use have rebounded, causing the death of beneficial organisms and the poisoning of the environment. DDT is now banned in many countries, and other effective but less persistent insecticides are being used.

8. Summary. From the foregoing sections it can be seen that we cannot regard the lives of the higher animals and plants as being divorced from other organisms of the earth.

There is a great diversity of species and groups and the higher animals and plants represent only a small percentage of the whole life on earth.

The impact of lower organisms on the higher seems often to be harmful as in the case of parasites and disease-causing organisms, and yet it should not be forgotten that many of the lower organisms are involved in the food chains from which we ultimately benefit and in the completion of the nutrient cycles upon which the growth of plants entirely depends.

PROGRESS TEST 11

1. What is the importance of the algae? (1)
2. Name a harmful protozoan. (2)
3. Why are most species of bacteria described as beneficial? (3)
4. What is an antibiotic? (3)
5. What is a saprophyte? (4)
6. How is alcohol made? (4)
7. What is a parasite? (5)
8. List the characteristics of an insect. (6)
9. Why is a ladybird considered to be a beneficial insect? (6)
10. What is a vector of disease? (6)

EXAMINATION TECHNIQUE

MOST biology examination questions do not have exact numerical answers and for this reason present problems both to the candidate and to the examiner. An attempt to examine biology more objectively is being made by a number of boards in the development of the *multiple-choice* type of question. In these the candidate is presented with a stem of information and a number of deductions that might be drawn from it and is asked to select the most valid item. While such questions have a good deal to recommend them (for example they can be marked mechanically) it does not seem to be the present policy of any board to use them for more than a part of the whole examination. It is therefore worth while to consider some ways in which conventional "essay style" or descriptive questions should be tackled.

1. Mark schemes. An important point to remember is that the examiner will have in front of him a mark scheme and that this will consist of facts exactly related to the question set. It is a fair assumption that this scheme will add up to twenty in the normal theory paper.

The first thing that the candidate must learn to do is to weigh up the importance of the sections in a question that consists of several parts. Thus, to take some examples:

"Indicate the major constituents of a balanced diet. How would the digestion of a meat sandwich occur?" (*Marking:* 6, 14.)

"How does a plant seed form? How does the seed become dispersed?" (*Marking:* 10, 10.)

Having assessed the relative importance of the parts of the question, and this is not difficult to do with a little practice, the candidate must weight his own answer accordingly.

The second thing to remember is that the mark scheme is precisely tailored to the question set and that answers must be completely relevant and factual. Vague generalisations and long rambling descriptions are useless. The mark scheme occupies only a few lines.

2. Allocation of time. Many candidates find the allocation of time under examination conditions very difficult, especially if they have been used to more or less unlimited time for essay writing. The majority of "O" Level biology papers are of 2 hours' duration so that only thirty-six minutes are available for any question (or thirty minutes in a two and a half hour paper). This means that a straightforward essay is not necessarily suitable.

Graphs, annotated diagrams, lists, sub-headings, flow diagrams are all ways in which a great deal of information can be put across in a short time. The underlining of key points is helpful to both examiner and candidate. Modern biological textbooks are not written in long uninterrupted sections nor is such a style suitable for examination answers.

A further point is that examination questions are not intended as vehicles for the candidate to show off irrelevant information. If a question on respiration states "Gaseous exchanges between lung and air" it means just this and therefore it is a waste of valuable time to include details of blood transport in an answer.

3. Problem questions. Candidates are advised to choose problem questions which have definite answers or other questions where exact information is required. The more general type of question is very difficult to answer and may be better avoided especially by mediocre candidates.

4. Practical examinations. As far as practical examinations are concerned several points are relevant. In the first place, many candidates fail to get down on paper the results of the time they have spent on a question. Remember that your script and perhaps drawing is all the examiner has and if you have not put the points across on paper about your observations and conclusions they will not be credited.

A large number of people are not artistic but anybody can draw diagrams of biological material that are large, neat and accurate. Credit is given for these three things, especially the latter.

5. Developing an examination technique. It is extremely important that candidates preparing for examinations should answer questions under the same sort of conditions that they will actually face at the time of the examination.

A recommended method is to carry out some preliminary study of a topic, look at a question on it, draw up a rough plan of your answer, then, concisely and factually answer the question from the head, not the book, in the time allowed.

EXAMPLES OF MULTI-CHOICE QUESTIONS

Complete this sheet as instructed at the beginning of each section (i), (ii) and (iii), and hand in the sheet with the rest of your answers.

(i) *Complete the following by writing the correct word in the box.*

The name of one parasite of man

Cells take in water by the process of

When plant cells are placed in water they become

Name two organs of a mammal from the list which are mainly concerned with excretion.   Salivary gland, oesophagus, lung, stomach, pancreas, kidney, rectum, sweat gland, anus.

The response of the root and shoot of a plant to gravity is called

Name the hormone whose deficiency leads to

(a)   cretinism

(b)   diabetes

A plant living entirely on the dead remains of other plants is called a

(ii) *Put one of the letters A, B, C, D, E, F, G, H, I in the box to indicate which of the terms you consider the most nearly correct answer.*

A photosynthesis, *B* translocation, *C* anaerobic respiration, *D* parasitism, *E* mutation, *F* pollination, *G* perennation, *H* reflex action, *I* fertilisation.

Movement of water and salts in a plant   ☐

Process producing alcohol   ☐

Spontaneous change in genetic make-up   ☐

Fusion of two gametes   ☐

Bright sunlight affecting the pupil size in the eye   ☐

(iii) *Assess each of the following statements by writing true or false in the box. Marks will be deducted for incorrect answers but not for boxes left blank.*

Sensory cells of the skin are receptors   ☐

Red blood corpuscles are destroyed only in the liver   ☐

All bacteria are parasites   ☐

Energy is lost from an organism during respiration   ☐

Diseases can be caused by bacteria, viruses and protozoa   ☐

Green plants always breathe out oxygen   ☐

(after O. & C.)

GLOSSARY OF TERMS

abdomen: the lower part of the trunk, containing the organs of digestion, excretion and reproduction.

absorption: the taking in of soluble substances across a cell membrane, also used in a wider sense, *i.e.* the absorption of food into the bloodstream.

acid: a sour substance which neutralises an alkali and turns litmus (a vegetable dye) red.

aerobic: the normal method of tissue respiration in which oxygen is used for the release of energy from food.

algae: a group of simple plants, mostly aquatic. Very important as the basic members of marine food chains.

alimentary canal: the digestive tract: a tube of specialised regions that runs from the mouth to the anus.

alkali: a substance which neutralises an acid and turns litmus (a vegetable dye) blue.

amino acid: the units from which proteins are built up and to which they are broken down before being taken into the bloodstream.

amnion: a membrane bag that surrounds the mammalian foetus and which is filled with water. It helps protect the foetus inside the uterus of the female during its period of development.

amylase: an enzyme that acts on the carbohydrate starch and breaks it down to a large number of malt-sugar molecules.

anaemia: a disease in which red blood corpuscles are deficient.

anaerobic: able to obtain energy from food without using oxygen.

anatomy: the study of the form and relationship of parts of an animal or plant.

antagonistic muscles: a pair of muscles that work in opposite directions across a joint; an example is the biceps and triceps of the human forearm.

antibiotic: a chemical substance (such as penicillin) isolated from an organism, which acts against bacteria and the other micro-organisms.

antibody: a chemical substance produced in the body to combat foreign matter such as bacteria.

antigen: substance in foreign bodies that stimulates the formation of antibodies.

antitoxin: a substance produced in the body to neutralise the toxins (poisons) produced by invading micro-organisms.

asexual reproduction: reproduction without sexual process.

automatic system: that part of the nervous system which controls visceral function, *e.g.* stomach.

bacteria: single-celled organisms, neither animal nor plant, but akin to the latter, some causing decay, some causing disease (*e.g.* typhoid).

bone: the supporting structure of the bodies of animals with back-bones, made up partly of organic material and partly of calcium phosphate.

brain parts: the brain is the controlling region of the central nervous system. In the front is the cerebrum, the main region of sensation and muscle control, while behind lies the cerebellum which controls balance. The brain stem at the top of the spinal cord is the medulla, and this is the centre of autonomic activity.

bronchioles: the smallest branches of the tubes that lead from the windpipe into the lungs and end in the air-sacs.

buccal cavity: the mouth cavity, leading to the windpipe and oesophagus.

buffering: taking acids or alkalis so as to make a neutral solution, as in the blood.

calories: the amount of heat required to raise the temperature of 1 gram of water 1 deg. C. One calorie = 4·2 joules.

capillaries: microscopic vessels connecting up arteries and veins, through the walls of which exchanges take place between blood and tissues.

carbohydrate: compound of carbon, hydrogen and oxygen, of which the latter two are in the ratio 2 : 1. The group includes sugars, starch, glycogen and cellulose.

cartilage: supporting tissue, softer than bone, found at the joints, in the lobe of the ear, and at the ends of the ribs.

cell: the smallest unit of a living organism. It has a controlling nucleus and surrounding cytoplasm. Plant cells have a dead cell wall and usually contain chloroplasts and vacuoles while animal cells have none of these features.

chlorophyll: the green substance in plants, which, with light energy, builds sugar from carbon dioxide and water.

chloroplasts: small structures, containing chlorophyll, in plant cells.

chromosomes: thread-like bodies in the nucleus, which carry genetic material.

cochlea: that part of the ear that turns sound into nerve impulses.

colon: the main part of the large intestine.

corpuscles: cells of various types and functions found in the blood.

cytoplasm: the protoplasm or living substance of a cell.

de-amination: the formation of ammonia from an amino acid preparatory to the manufacture of urea in the liver.

deficiency disease: a condition due to a lack of, or failure to use, a particular item of diet.

dehydrated: having lost a lot of water.

DNA: (deoxyribose nucleic acid), the substance of the gene, a long molecule which can incorporate genetic instructions.

diaphragm: the sheet of muscle separating thorax and abdomen; it is important in breathing.

diffuse: to spread out so as to occupy a space or liquid evenly; in the body this is frequently through a membrane.

digestion: the breaking down of complex foods in the alimentary canal so that they may be absorbed into the blood.

disaccharide: a carbohydrate made up of two sugar units, *e.g.* cane sugar.

disease: a departure from normal health.

dominant gene: a gene that shows itself when combined in a zygote with a contrasting gene.

duodenum: the first part of the small intestine, between stomach and ileum.

effector: an organ or system reacting to a stimulus.

embryo: a mammal in its developmental stage which takes place within the uterus of the female.

emulsify: to break up into minute droplets; *e.g.* fat is emulsified by bile.

endocrine gland: a ductless gland (*e.g.* pituitary or thyroid gland) which produces a hormone. The product passes directly into the blood.

endoplasmic reticulum: the network within the cytoplasm of cells on which synthesis of the cells, chemicals and secretions takes place.

energy: the power to do work; this may be as heat, electrical, chemical or mechanical energy. Now measured in joules.

environment: the surroundings in which a plant or animal lives.

enzyme: a chemical (protein) found in living cells, which assists some particular reaction to take place. Enzymes are sensitive to heat and are destroyed at high temperatures.

epidermis: the outer layers of the skin.

evolution: the establishment of species of living things by gradual change from simple forms through many generations.

excrete: to eliminate the waste products of bodily functions.

faeces: waste matter from the food tract, excreted at the anus.

fat: the combination of a fatty acid and glycerol, a necessary part of the balanced diet.

foetus: the immature stage of a mammal before birth.

follicle: a group of cells from which a special development starts (*e.g.* the hair follicle).

fungi: plants without chlorophyll that obtain nourishment by breaking down and absorbing the organic materials of other living or dead organisms.

gamete: a sex cell; in higher animals it is normally found as a sperm from the male or an egg from the female.

gastric: to do with the stomach; *e.g.* the gastric juice is the digestive juice formed in the stomach.

gene: the chemical unit on the chromosome of a cell nucleus that determines some special character (*e.g.* eye colour).

germination: the process whereby a seed takes in water and gives rise to a root and shoot to form a new independent plant.

glycerol: part of the fat molecule; it unites with different fatty acids to give different fats.

glycogen: (also known as animal starch). It consists of many sugar units and is found in liver and muscles as storage carbohydrate.

haemoglobin: the pigment of red blood corpuscles; it is responsible for the carriage of oxygen.

homozygous: an organism bearing two similar genes for a given character. The opposite is *heterozygous* where the genes carried are for contrasting characters.

hormone: a secretion produced in an endocrine (ductless) gland and having a specific effect (*e.g.* sexual maturation or growth).

hydrolysis: the breaking down of molecules into smaller ones with the addition of water. This process plays an important part in digestion.

ileum: the lower part of the small intestine; it is mainly concerned with absorption of digested food into blood and lymph.

ilium: the main part of the pelvic girdle.

ions: the soluble state of an element which is formed in water and can diffuse across a membrane. Sometimes ions are combinations of the element with hydrogen or oxygen. Ions carry positive or negative charges.

lacteal: a vessel of the lymph system.

lamina: the blade of a leaf.

ligament: the connective tissue joining one bone to another.

lipase: a fat-splitting enzyme, as in pancreatic juice.

lymph: a colourless fluid derived from blood; it serves as a go-between for blood and tissues.

mammal: a warm-blooded backboned animal, the female of which suckles its young; *e.g.* a human being.

maturation: the process of development to a final working state; thus an egg or sperm can come to maturation and so can a whole animal.

medulla: the centre of an organ or tissue.

meiosis: making of sex cells by halving the chromosome number and exchanging genetic material.

membrane: the thin layer bounding the protoplasm of a cell; also any thin sheet of tissue.

meristem: a part of a plant which can propagate new cells, normally the root and stem tip and the cambium.

mesophyll: the central cells in leaves.

metabolism: the chemical changes of building up and breaking down in a living organism.

mitochondria: cell parts involved in energy release from food.

mitosis: division of a cell into two similar daughter cells.

molecule: a chemical combination of atoms of elements; *e.g.* a molecule of oxygen is O_2.

monosaccharide: the basic sugar unit (such as glucose $C_6H_{12}O_6$) from which all more complex carbohydrates are built up, and to which they are broken down in digestion.

mucus: a slimy substance, produced by cells in the inner lining of body tracts (*e.g.* the gut), which acts as a lubricant.

muscle: animal tissue specialised for contraction, by which movements are brought about.

nitrogenous: containing nitrogen; *e.g.* substances such as proteins or urea.

non-permeable: used of layers, such as membranes, which do not allow free passage of substances.

nucleus: that part of the cell which controls its activity and is important in cell reproduction.

nutrient: a substance of food value to an organism.

nutrition: the taking in and utilisation of food.

oesophagus: the tube between the buccal cavity and the stomach.

olfactory: to do with the sense of smell.

omnivore: an animal that eats both plant and animal food. Man is a good example of an omnivore.

organic compound: chemical substance containing carbon and usually hydrogen and oxygen. Many such compounds occur in protoplasm.

organism: a living thing.

organs: groupings of specialised tissues forming a particular function; *e.g.* the heart.

osmo-regulation: the process of water control to maintain the correct proportion of liquid in the body.

osmosis: the tendency of water to pass across a membrane from a weaker to a stronger solution, thus setting up osmotic pressure.

oxidation: the union of oxygen with chemical substances, usually with the release of energy, as in burning. Sugar is oxidised in the body for this purpose.

parasite: an organism living upon another, deriving food from it and causing harm; *e.g.* malaria is a parasite on man.

pathogenic: causing disease.

peptone: product of the first stage of protein digestion.

perennation: survival of plants over winter.

peristalsis: a type of alternating contraction and relaxation found in the gut; it ensures the onward passage of food.

permeable: used of a layer or membrane which allows the passage of substances.

phagocyte: a white blood corpuscle able to destroy bacteria and other foreign bodies.

phloem: the vascular tissue which conducts organic substances in plants.

photosynthesis: the fixation of carbon dioxide by plants to form sugar and other substances. It involves the action of chlorophyll and the provision of light energy.

physiology: the study of the functions of living things.

placenta: the link between the uterus and the embryo in the mammal.

pollination: the transfer of pollen from the anther to the stigma in plants of the same species.

polypeptide: a stage in protein digestion.

polysaccharide: a carbohydrate built up of many sugar units (saccharides); *e.g.* starch.

proteins: substances fundamental in all living matter, containing nitrogen, carbon, hydrogen, oxygen and often other elements.

pulmonary: to do with the lungs; *e.g.* pulmonary artery and vein.

receptor: a sense organ of the body specialised to respond to a particular change in the surroundings, *e.g.* temperature, light or sound.

reflex: the automatic response to a nervous stimulation.

rennin: a milk-coagulating enzyme in the stomach of young animals.

respiration: (1) external respiration is breathing; (2) internal or tissue respiration is the chemical breakdown of food in the tissues to produce energy.

response: an act carried out by an organ or the whole body as the result of something (a stimulus) that has been done to it.

RNA: (ribose nucleic acid) carries the instructions from the nucleus to the rest of the cell. Important in protein synthesis.

saprophytes: organisms living on the products of the remains of plants and animals, and having no chlorophyll of its own; *e.g.* fungi and many bacteria.

sebaceous gland: a grease gland in the skin.

secretion: (1) the production and giving out of materials or juices by living cells or glands; (2) the product so formed.

sense organ: a specialised part of the body of an organism which can detect changes taking place in the environment; *e.g.* the eye.

sexual reproduction: the production of a new individual from the fusion of sperm (male) and female (egg) sex cell.

spore: the resistant body formed by bacteria and other simple organisms to ensure survival or as a reproductive method.

stimulus: a change in the surroundings of an organism to which it will respond. Stimuli can also be internal; *e.g.* increasing carbon dioxide in the blood acts as a stimulus which causes the lungs to work faster.

stomata: small pores in leaves which allow diffusion of gases.

synovial fluid: the lubricating liquid found in certain movable joints of the skeleton.

tendon: tissue that connects a muscle to a bone.

thorax: the chest cavity and region of the body.

tissue: collection of similar cells specialised for particular functions; *e.g.* nerves.

toxin: a poison; the word is usually used for the poisons produced by bacteria.

trachea: the windpipe leading to bronchi and lungs.

translocation: the movement of organic matter through plants.

transpiration: the movement of water through a plant.

tropism: movement of part of a plant towards or away from a stimulus.

tubule: a minute tube, as in the filter unit of a kidney.

urea: the chief end-product of protein breakdown in the liver, of formula $CO(NH_2)_2$.

ureter: the tube connecting the kidneys to the bladder.

urethra: the tube from the bladder to the outside of the body, through which urine passes.

vacuole: a minute hole in protoplasm, containing watery solution.

vector: a carrier of disease.

vegetative reproduction: increase by non-sexual means, usually referring to plants.

villi: minute finger-like projections of the small intestine, which absorb food.

virus: an extremely small organism on the borders of the living and the non-living. It is visible only with the electron microscope. Many diseases (*e.g.* poliomyelitis) are caused by viruses.

vitamin: a constituent of the food needed in small amounts to ensure a certain aspect of health.

zygote: an egg cell after fertilisation by a sperm.

INDEX

abscission layer, 39
acidity/alkalinity, 88
adenosine triphosphate (ATP), 2, 48, 73–4, 82–3
adrenalin, 70, 115
adrenals, 115
aerobic respiration, 2, 30, 74
agricultural revolution, 101
alcohol, 2, 75, 147
algae, 28, 142
alveoli, 76–8
amino-acid, 20
amnion, 135–6
amylase, 64, 66
anaerobic respiration, 2, 74
analysis of living matter, 16
anaphase, 123, 125
animals and plants compared, 14
annual growth, 37
antagonistic muscles, 117–18
anther, 128
antibiotic, 145, 147
antibody, 95
antigens, 95
antitoxins, 95
aorta, 79–80
arteries, 80
ascorbic acid, 56
asexual reproduction, 7
assimilation, 68–9
auricle, 79
autonomic system, 114, 119
autotropic, 54
auxin, 34, 108
axillary buds, 38

bacteria, 3, 61, 65, 70, 94, 100, 144–146
balance, 111
balanced diet, 54–5, 59
bark, 36, 38
bases, 21
bees, 128, 149

beriberi, 56
biceps, 117
bile, 66
birth, 136
blood, 79–80
brain, 113–14
buccal cavity, 64

cambium, 35–6, 38
canines, 61
carbohydrates, 17–18, 55
carbon cycle, 97
carbon dioxide, 48, 50, 78, 84, 89, 97–8, 99
carbon isotope, 48
carnivores, 62
carpel, 128
cell division, 122 ff
    membrane, 11
    walls, 12
cells, electron microscope and, 10
    generalised, 9
    plant, 11–12
cellulose, 12, 18, 49
cerebellum, 113
cerebrum, 113
chemotaxis, 133
Chlamydomonas, 142
chlorophyll, 46
chloroplast, 11
choroid, 110
chromosome, 123, 133
clinostat, 107
coccus, 145
condensation, 19, 20
cones, 109
cork, 34, 38
cornea, 108
corpuscles, 81
cotyledon, 129
crossing over, 124
cuticle, 39